Jianguo Li · Jianxiong Pan · Yujie Lin · Neng Ye · Kai Yang

Key Technologies of High Frequency Wireless Communications

 Springer

Jianguo Li
School of Cyberspace Science
and Technology
Beijing Institute of Technology
Beijing, China

Yujie Lin
Beijing Institute of Technology
Beijing, China

Kai Yang
School of Information and Electronics
Beijing Institute of Technology
Beijing, China

Jianxiong Pan
Beijing Institute of Technology
Beijing, China

Neng Ye
Beijing Institute of Technology
Beijing, China

ISBN 978-981-96-5893-0   ISBN 978-981-96-5894-7   (eBook)
https://doi.org/10.1007/978-981-96-5894-7

This Springer imprint is published by the registered company Springer Nature Singapore Pte Ltd.
The registered company address is: 152 Beach Road, #21-01/04 Gateway East, Singapore 189721,
Singapore

If disposing of this product, please recycle the paper.

Key Technologies of High Frequency Wireless Communications

Preface

Wireless communication systems operating in high-frequency bands, particularly millimeter wave communication and terahertz communication, have emerged as pivotal broadband access solutions for 6th generation mobile communication network (6G). While promising terabit-per-second transmission rates, microsecond-scale latency thresholds, robust physical-layer security, and integrated communication-sensing hardware architectures, these high-frequency systems confront intrinsic challenges stemming from significant path attenuation and constrained hardware capabilities. The goal of this book is to offer a comprehensive exploration of high-frequency wireless communication technologies. It focuses on different aspects such as high-spectrum efficiency transmission technology, secure directional modulation technology, and integrated technology of high-frequency wireless communications and high-precision ranging. For each aspect, relevant techniques and algorithms are discussed and analyzed to improve performance and address challenges. In addition, this book has provided elaborate simulation results for each topic to verify the feasibility of the corresponding schemes.

In particular, for high-frequency and long-distance communication, we discuss an efficient system: multiple-input multiple-output (MIMO) system with full-duplex device-to-device (D2D) communications, show an efficient detection algorithm: asynchronous multi-user detection for code-domain non-orthogonal multiple access (NOMA), and present an efficient coding and decoding algorithm: quasi-cyclic low-density-parity-check (LDPC) transceiving framework. For high-frequency physical layer secure communication, we investigate multi-user hybrid beamforming design for line-of-sight (LOS) communications and the incorporation of secure directional modulation in reconfigurable intelligent surface (RIS)-aided communication networks. For the constrained resources in high-frequency communication, we further explore the various integrated waveform designs, including orthogonal frequency division multiplexing (OFDM) Radar and advanced discrete Fourier transformation spreading-based OFDM (DFT-s-OFDM).

We believe that the results in this book can provide useful insights into the design of high-frequency wireless communication systems in the future. This book is intended for graduate students, researchers, and engineers in the field of wireless communications. We hope that the contents of this book can provide useful guidance to the readers.

Beijing, China   Jianguo Li
                                                                           Jianxiong Pan
                                                                               Yujie Lin
                                                                                Neng Ye
                                                                               Kai Yang

Competing Interests The authors have no competing interests to declare that are relevant to the content of this manuscript.

Contents

Acronyms

3D-EPA	Three-Dimensional-Expectation Propagation Algorithm
3GPP	The 3rd Generation Partnership Project
5G	5th Generation mobile communication network
5G NR	5G New Radio
6G	6th Generation mobile communication network
AMP	Approximate message passing
ATSC	Advanced Television Systems Committee
AWG	Arbitrary waveform generator
AWGN	Additive White Gaussian Noise
BER	Bit Error Rate
BP	Belief Propagation
BP-EGC-MPA	Belief Propagation with equal gain combining message passing algorithm
BP-MPA	Belief Propagation message passing algorithm
BPSK	Binary phase shift keying
BRAM	Block RAM
BS	Base station
CCSDS	Consultative Committee for Space Data Systems
CE	Cross-Entropy
CE-OFDM	Constant envelope orthogonal frequency-division multiplexing
CNC	Computer numerical control
CNP	Check node processing
CNU	Check node processing unit
CP	Cyclic prefix
D2D	Device-to-device
DAC	Digital to analog conversion module
DFT	Discrete Fourier transform
DFT-s-OFDM	Discrete Fourier transformation spreading-based OFDM
DNN	Deep neural networks
DSP	Digital signal process
DVB	Digital video broadcasting

EGF	Extended Gaussian Filter
EM	Expectation maximization
eMBMS	Evolved multimedia broadcast multicast services
EPA	Expectation propagation algorithm
ESE	Elementary signal estimation
FBMC	Filter bank multicarrier
FCC	Federal Communications Commission
FDE	Frequency-domain equalization
FDM	Frequency division multiplexing
FDMA	Frequency division multiple access
FDSS	Frequency domain spectral shaping
FFT	Fast Fourier Transform
FMCW	Frequency modulated continuous wave
FN	Function node
FPGA	Field-programmable gate array
FRFT	Fractional Fourier transform
FTN	Faster-than-Nyquist
HUE	Hardware utilization efficiency
ICI	Inter-channel interference
IDFT	Inverse Discrete Fourier Transform
IFFT	Inverse Fast Fourier Transform
IOTA	Isotropic orthogonal transform algorithm
ISAC	Integrated sensing and communications
ISI	Inter-symbol interference
ITU	International Telecommunication Union
IUI	Inter-user interference
K-L	Kullback-Leibler
LAN	Local area network
LDM	Layered division multiplexing
LDPC	Low-density parity check
LL	Lower layer
LLR	Log-likelihood ratio
LLR-BP	Likelihood ratios BP algorithm
LO	Local oscillator
LOS	Line of sight
LTE	Long Time Evolution
LUT	Look-Up Tables
MCSs	Modulation coding schemes
MIMO	Multiple-input multiple-output
MMSE	Minimum mean square error
MMSE-SIC	Minimum mean square error-successive interference cancellation
mmWave	Millimeter wave
MPA	Message passing algorithm
MRT	Maximum ratio transmit
MSA	Min-Sum Algorithm

NLOS	Non-Line of Sight
NMS	Normalized Min-Sum
NOMA	Non-orthogonal multiple access
NOW	Non-orthogonal waveform
OFDM	Orthogonal frequency division multiplexing
OOB	Out of band
P/S	Parallel into serial
PAPR	Peak-to-average power ratio
PFN	Prior factor node
PLS	Physical layer security
QC-LDPC	Quasi-cyclic-low density parity check
QoS	Quality of service
RAM	Random Access Memory
RCS	Radar cross-section
RF	Radio frequency
RIS	Reconfigurable intelligent surface
RMSE	Root mean square error
RRC	Root-raised cosine
SCMA	Sparse code multiple access
SCS	Subcarrier spacing
SE	Spectrum efficiency
SEFDM	Spectrally efficient frequency-division multiplexing
SINR	Signal-to-interference-plus-noise ratio
SNR	Signal-to-noise ratio
TDM	Time division multiplexing
THz	Terahertz
T-SIC	Triangle SIC
TU-6	Typical Urban 6 path
UL	Upper layer
VN	Variable node
VNP	Variable node processing
VNU	Variable node processing unit
WBE	Welch-bound equality
WiGig	Wireless Gigabit Alliance
XR	Extended reality
ZC	Zadoff–Chu
ZF	Zero forcing

Chapter 1
Introduction

This chapter first explains the research necessity of high-frequency wireless communications. Then the development of high-frequency wireless communication systems is reviewed, and the shortcomings of the existing research are summarized from the physical layer transmission level combined with the current background. Finally, the organizational structure of the book is introduced.

1.1 Background

With the exponential growth of wireless communication traffic, it becomes challenging for low-frequency wireless transmission technology to meet users' demands for ultra-high speed, low latency, and high quality of service. High-frequency bands such as mmWave and terahertz (THz) offer wide bandwidth and large system capacity, enabling the realization of ultra-high-speed wireless communications. In October 2012, the Wireless Gigabit Alliance (WiGig) approved the release of IEEE 802.11ad standard primarily used for hotspots and wireless local area network (LAN) coverage with a carrier band at 60 GHz. In October 2015, the International Telecommunication Union (ITU) officially announced eight candidate bands ranging from 24.25 GHz to 86 GHz for 5th generation mobile communication network (5G) deployment. In July 2016, the Federal Communications Commission (FCC) released four mmWave bands from 27.5 GHz to 71 GHz with a total bandwidth of 11 GHz. In March 2019, the FCC decided to open up the THz band to encourage research on THz wireless communications in order to facilitate its early adoption in future generation services like 6G. Researchers have extensively investigated high-frequency and high-speed wireless communications, as depicted in Table 1.1 [1–10].

As depicted in Table 1.1, the frequency band has gradually evolved from E-band and 120 GHz to higher frequency bands such as 220 GHz and 340 GHz. In terms of communication speed, it has progressed from less than 10 Gbps to more than

© The Author(s), under exclusive license to Springer Nature Singapore Pte Ltd. 2025

J. Li et al., *Key Technologies of High Frequency Wireless Communications*,

https://doi.org/10.1007/978-981-96-5894-7_1

Table 1.1 Comparison of high-frequency wireless communication systems

Frequency (GHz)	Speed (Gbps)	Distance (km)	Modulation	References
73	5	–	16QAM	[1]
73.5	8	–	16QAM	[2]
83	9/16	–	64QAM/QPSK	[3]
120	22.2	–	QPSK	[4]
140	5	21	16QAM	[5]
143	5.3	0.01	64QAM	[6]
220	3.5	0.2	QPSK	[7]
300	42	–	2ASK	[8]
340	3	0.05	16QAM	[9]
350	100	0.002	16QAM	[10]

100 Gbps. Regarding transmission distance, there has been a gradual shift from laboratory waveguide connection verification to point-to-point communication verification spanning tens of kilometers. In the field of modulation and demodulation, the relatively simple low-order amplitude modulation with low spectral efficiency has advanced towards complex high-order modulation techniques with improved spectral efficiency. In terms of implementation methods, there has been a transition from a photoelectric combination or analog mode to an all-digital modulation and demodulation scheme. The all-digital modulation and demodulation scheme offers a high spectrum utilization rate and strong anti-interference performance, enabling high-speed and long-distance transmission. Figure 1.1 illustrates the transmission rate and modulation order for single-channel, single-carrier, single-polarization, and high-speed communication systems in published high-frequency wireless communication systems.

As depicted in Fig. 1.1, the majority of communication systems operating at a single-channel transmission rate exceeding 50 Gbps employ offline demodulation or same-source transceiver mode [11–26], which is impractical for real deployment. In scenarios with high transmission rates, low-order modulation schemes are employed to ensure wider bandwidth and simpler communication transceiver structures with reduced complexity. However, when employing high-order modulation techniques, single-channel communication systems impose stringent hardware requirements such as power amplifiers and high-speed analog-to-digital converters, making it challenging to achieve ultra-high-speed wireless communications. Most existing communication schemes currently support a maximum communication rate of 25 Gbps.

In addition, high-frequency wireless transmission encounters numerous challenges arising from short wavelengths, significant propagation losses and weak devices. Firstly, the propagation loss of high-frequency signals is large, resulting in poor coverage ability. Secondly, due to constraints in resources, volume, and power consumption, the limitations of traditional encryption and decryption are becoming increasingly prominent, challenging to achieve physical layer security. Furthermore,

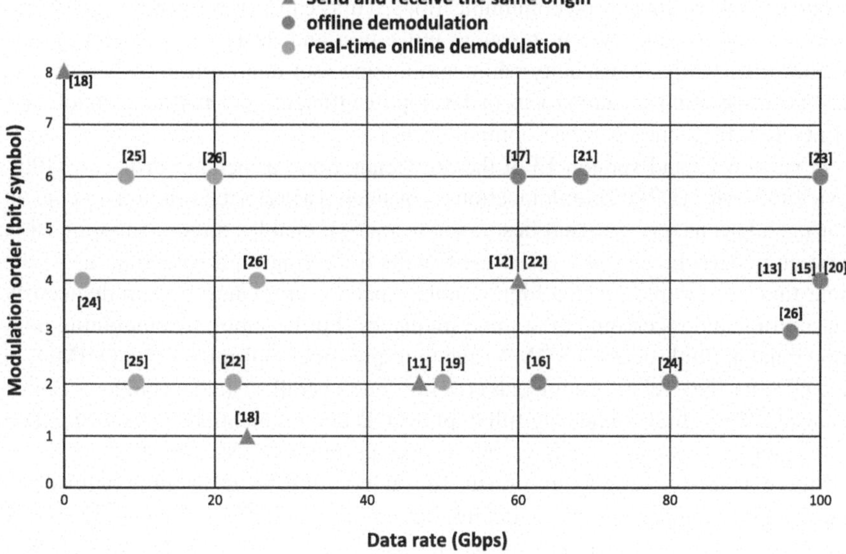

Fig. 1.1 Relationship between transmission rate and modulation order of existing high-frequency and high-speed communication systems

the existing separate design of communication and sensing with independent functions and high information processing delay leads to a waste of spectrum resources and hardware resources, which is not beneficial to lightweight design. Thus, high-frequency wireless communications put forward higher requirements for spectral efficiency, communication security, and hardware efficiency.

1.2 High-Spectral-Efficiency Transmission in High-Frequency Communication

The primary drawbacks of high-frequency transmission include significant signal propagation loss and limited coverage, which poses challenges for meeting the demands of high-rate and multi-user access in 6th generation mobile communication network (6G). Multi-antenna technology, multi-access methods, terminal direct connection technology, efficient coding technology, and their integrated application can enhance spectral efficiency, expand coverage area, and increase system capacity. Current high-frequency communication systems adopt a direct coverage mode with multi-antenna base stations that have limited range, low system capacity, and low spectral efficiency for cell edge users. Meanwhile, terminal direct connection technology enables user-to-user communication without passing through the base station, reducing pressure on the base station while decreasing transmission delay and increasing cell coverage. However, since it uses the same time-frequency resource block as other users' communications, it may cause interference to cell users. Therefore, optimizing system resource utilization and designing efficient system architecture to improve capacity while minimizing user interruption probability has become an essential research focus in developing efficient transmission technologies for 6G high-frequency wireless communications.

In version 12 and version 13 of the 3rd Generation Partnership Project (3GPP), device-to-device (D2D) communication is considered a preferred solution for content sharing, adjacent services, and business coverage. It enables direct communication between geographically proximate users without relying on a base station. At the same time, full-duplex technology doubles spectral efficiency by simultaneously transmitting and receiving at the same frequency. Furthermore, by combining non-orthogonal multiple access (NOMA) and multiple-input multiple-output (MIMO) in the D2D auxiliary relay system, cell coverage and overall system throughput can be enhanced. Thus, high-frequency full duplex D2D communication systems based on non-orthogonal transmission have emerged as a prominent research area.

Most existing research focuses on the fusion level of heterogeneous networks, primarily addressing power control and interference coordination strategies in network nodes. At the physical layer transmission, D2D communication technology enables spectrum resource reuse within cells. Nevertheless, current half-duplex D2D communication exhibits low efficiency, while full-duplex D2D communication poses significant challenges due to self-interference. Currently, there is limited research on performance analysis and deployment recommendations for full-duplex D2D in high-frequency wireless communication systems.

1.3   Array Security Modulation Technology in High-Frequency Communication

The security of high-frequency communication is also a key focus for the future 6G. With the number of access nodes increasing, the privacy and security of transmitted data are more susceptible to threat. The increase in wireless connections in 6G, the large number of delay-sensitive devices, and the limited hardware resources of terminal devices put forward higher requirements for communication security. The encryption of traditional communication systems often adopts a cryptographic scheme. However, the confidentiality performance of this scheme gradually deteriorates with the significant improvement of the computing performance of eavesdroppers. Nevertheless, secure transmission at the physical layer can utilize the randomness and uniqueness of the wireless channel to efficiently achieve the secure transmission of information without occupying additional communication time and computational resource overhead of the system. The wavelength of high-frequency band wireless communication systems is short, and the antenna array scale is large, which can efficiently implement directional modulation technology. Specifically, directional modulation can send a standard constellation diagram in the direction of the target user and a noise-like constellation diagram in the direction of the eavesdropping user to achieve secure transmission at the physical layer. However, multi-user directional modulation will lead to the superposition of sidelobe energy, poor beam focusing, and low transmission efficiency. Therefore, it is urgent to study the directional modulation technology with high-frequency, multi-user, low-sidelobe and security characteristics for 6G.

However, most of the existing researches focus on optimizing physical layer security performance such as secure transmission rate, interruption probability, and secure capacity through system beamforming design. These methods can achieve optimal security, but they rely on the channel environment of eavesdropping users being worse than that of target users. Nevertheless, the direction modulation technology can randomize the constellation map received by eavesdropping users. Current research mainly focuses on a single secure user. Additionally, the sending side lobe energy is high, leading to low transmission efficiency.

1.4 Integration of Communication and Ranging in High-Frequency Systems

The discrete design of high-frequency communication and perception serves independent functions, resulting in high information processing delay and wastage of spectrum and hardware resources. In the future, 6G's requirements for communication and perception will be coupled with each other. On the one hand, the massive increase in sensing service measurement data necessitates high-frequency broadband transmission to meet its data rate demands. On the other hand, real-time perception of the environment by the high-frequency communication system can enhance communication performance and enable intelligent connectivity. High-frequency signals possess characteristics such as short wavelength, high resolution, and antenna integration feasibility that facilitate wireless sensing services like precise ranging and angle measurement. However, existing discrete designs for communication and ranging suffer from issues such as spectrum waste, resource inefficiency, elevated costs, independent functionality, and substantial information processing delays. Considering the design requirements of a 6G multi-functional integrated waveform, it is necessary to use hardware and software resource sharing or information sharing to achieve efficient and intelligent transmission of the system, and it is necessary to take into account communication performance and measurement accuracy, and optimize for different application scenarios. Consequently, there is an urgent need to develop an integrated communication and ranging system with heightened precision yet low complexity tailored explicitly for 6G networks to realize joint resource scheduling capabilities, thereby enhancing overall performance encompassing both communication aspects as well as perception functionalities.

Due to the high frequency of mmWave and THz signals, the high-frequency communication system can achieve precise ranging while ensuring safe and effective transmission. Linear frequency modulation signal radar can achieve measurement accuracy at the micrometer level in the mmWave, while laser interferometers can achieve microsecond level measurement accuracy under very narrow bandwidth conditions, albeit with limited range. While orthogonal frequency division multiplexing (OFDM) radar in mmWave enables high-precision ranging during communication. Currently, the research primarily focuses on the integrated design of communication-ranging waveforms. In waveform design, ranging mainly considers target parameter resolution, maximum detection distance, and maximum range without ambiguity, whereas communication primarily considers communication rate and bit error rate (BER), both of which have opposing yet unified requirements for system parameters. The existing OFDM waveforms compromise ambiguity function performance by deteriorating the quality of communicated information carried within them, reducing the coherent accumulation performance of target echoes and estimation accuracy of target parameters.

1.5 Organization

To have a comprehensive view of high-frequency wireless communication technologies, this book discusses high-spectrum-efficiency transmission technology from three perspectives, namely system, detection and coding in Chaps. 2 to 4, presents the high-frequency secure directional modulation technology in Chaps. 5 and 6, and explores the integrated technology of high-frequency wireless communications and high-precision ranging in Chaps. 7 to 9. The organization of the rest of this book is as follows.

Chapter 2 investigates a full-duplex D2D-aided mmWave MIMO-NOMA broadcasting system. A layered division multiplexing (LDM) structure and full-duplex D2D communication are applied to achieve better performance, which are also the key techniques used in this system. In this chapter, the outage probability and the ergodic capacity are derived by using the Laplace transform technique and then used to analyze the system performance. The system shows better performance compared to conventional time division multiplexing and LDM without D2D communications.

Chapter 3 focuses on asynchronous code-domain NOMA detection. In scenarios with high channel dynamics or simplified scheduling, synchronization becomes difficult, leading to complex inter-user interference (IUI). A three-dimensional-expectation propagation algorithm (3D-EPA) for low-complexity detection is introduced, which iterates between per-user and inter-user steps and is extended to multi-antenna scenarios with state evolution analysis. Compared to other methods, it has a better BER performance, especially under high overloadings, thereby enhancing spectral efficiency in high-frequency wireless communications.

Chapter 4 centers around the quasi-cyclic-low density parity check (QC-LDPC) transceiving system, a combined design of a novel low density parity check (LDPC) code structure and the corresponding overlapping decoding strategies. For the code structure, a consultative committee for space data systems (CCSDS)-like quasi-cyclic parity check matrix with uniformly distributed submatrices is constructed to maximize overlap depth for parallel decoding. In terms of the decoding strategy, a modified

2-bit Min-Sum algorithm is used, achieving a 5 dB coding gain at a bit error rate of 10^{-6} compared to uncoded binary phase shift keying (BPSK) and reducing resource consumption. Simulation and implementation results show that the decoder can reach a throughput of 7.76 Gbps at 156.25 MHz with eight iterations and save two-thirds of resource consumption, thus improving the performance of high-frequency wireless communication systems.

Chapter 5 introduces a multi-user hybrid beamforming design for a multi-user physical layer security modulation technique. The beamforming scheme is utilized in the base station to generate multi-beams according to the direction angle of the target users. A cross-entropy iteration method is proposed to choose the optimal antenna combination, aiming to reduce the sidelobe energy. The performance of this method is analyzed in terms of sidelobe energy and symbol error rate, demonstrating its superiority over some traditional methods in enhancing the physical layer security of the system.

Chapter 6 further elaborates on secure directional modulation in reconfigurable intelligent surface (RIS)-aided networks with a low-sidelobe hybrid beamforming approach. It considers the incorporation of secure directional modulation in RIS-aided communication networks and proposes a novel cross-entropy iterative method to achieve low-sidelobe hybrid beamforming. The chapter uses maximum sidelobe energy as the objective function and deploys Kullback-Leibler divergence as the criterion in the iterative selection of good candidate antenna subsets. Through simulations, the performance of the proposed scheme is evaluated, showing its effectiveness in reducing sidelobe energy and maintaining a certain symbol error rate for eavesdroppers, thus ensuring physical-layer security.

Chapter 7 explores the integration technology of communication and sensing. The technology aims to achieve a more efficient and intelligent communication system by integrating sensing capabilities into the communication process. This chapter analyzes the key challenges and opportunities in this integration, and proposes novel algorithms and architectures to optimize the performance of the integrated system. Compared to traditional communication or sensing systems, the proposed techniques in this chapter can achieve better performance in terms of accuracy, efficiency, and resource utilization, enabling more reliable and intelligent wireless communication services.

Chapter 8 investigates the enhancement of peak-to-average power ratio (PAPR) and throughput for discrete Fourier transformation spreading-based OFDM (DFT-s-OFDM) system. Faster-than-Nyquist (FTN) signaling is integrated into the DFT-s-OFDM to enhance spectral efficiency, while an isotropic orthogonal transform algorithm (IOTA) filter with good time-frequency focusing ability is deployed to reduce the inter-symbol interference (ISI) introduced by FTN. Simulation results show that the proposed waveform achieves 3.5 dB PAPR gain and 50% throughput gain compared to conventional waveforms, demonstrating its effectiveness in enhancing the performance of high-frequency wireless communication systems.

Chapter 9 presents an ISAC waveform that adopts frequency domain spectral shaping (FDSS) to enhance the performances of DFT-s-OFDM, including sensing accuracy and PAPR, by adjusting the correlation of signals. The pre-equalization filter is used to equalize the amplitude of the DFT-s-OFDM signal, making the frequency domain signal amplitude more uniform. Simulation results indicate that the proposed scheme can obtain about 4 dB performance gain in terms of sensing accuracy over DFT-s-OFDM, significantly reduce PAPR, and improve the power amplifier efficiency, verifying its effectiveness in enhancing the performance of 6G wireless communications and sensing systems in high-frequency bands.

Chapter 10 summarizes this book and discusses the future directions of high-frequency wireless communications.

References

1. Z. He, J. Chen, C. Svensson, L. Bao, A. Rhodin, Y. Li, J. An, H. Zirath, A hardware efficient implementation of a digital baseband receiver for high-capacity millimeter-wave radios. IEEE Trans. Microwave Theory Techniques **63**(5), 1683–1692 (2015)
2. P. Harati, A. Dyskin, and I. Kallfass, Analog carrier recovery for broadband wireless communication links, in *2018 48th European Microwave Conference (EuMC)*, pp. 1401–1404 (2018)
3. S. An, Z.S. He, J. Chen, H. Han, J. An, H. Zirath, A synchronous baseband receiver for high-data-rate millimeter-wave communication systems. IEEE Microwave Wirel. Components Lett. **29**(6), 412–414 (2019)
4. H. Takahashi, A. Hirata, J. Takeuchi, N. Kukutsu, T. Kosugi, and K. Murata, 120-GHz-band 20-Gbit/s transmitter and receiver MMICs using quadrature phase shift keying, in *2012 7th European Microwave Integrated Circuit Conference*, pp. 313–316 (2012)
5. Q. Wu, C. Lin, B. Lu, L. Miao, X. Hao, Z. Wang, Y. Jiang, W. Lei, X. Den, H. Chen, J. Yao, J. Zhang, A 21 km 5 Gbps real time wireless communication system at 0.14 THz, in *2017 42nd International Conference on Infrared, Millimeter, and Terahertz Waves (IRMMW-THz)*, pp. 1–2 (2017)
6. V. Vassilev, Z.S. He, S. Carpenter, H. Zirath, Y. Yan, A. Hassona, M. Bao, T. Emanuelsson, J. Chen, M. Hörberg, Y. Li, J. Hansrydl, Spectrum efficient D-band communication link for real-time multi-gigabit wireless transmission, in *2018 IEEE/MTT-S International Microwave Symposium - IMS*, pp. 1523–1526 (2018)
7. Z. Chen, B. Zhang, Y. Zhang, G. Yue, Y. Fan, Y. Yuan, 220 GHz outdoor wireless communication system based on a Schottky-diode transceiver. Ieice Electron. Express **13**(9), 20 160 282–20 160 282 (2016)
8. Y. Takahashi, S. Noda, Breakthroughs in photonics 2013: a microwatt-threshold Raman silicon laser. IEEE Photonics J. **6**(2), 1–5 (2014)
9. C. Wang, B. Lu, C. Lin, Q. Chen, L. Miao, X. Deng, J. Zhang, 0.34-THz wireless link based on high-order modulation for future wireless local area network applications. IEEE Trans. Terahertz Sci. Technol. **4**(1), 75–85 (2014)
10. K. Liu, S. Jia, S. Wang, X. Pang, W. Li, S. Zheng, H. Chi, X. Jin, X. Zhang, X. Yu, 100 Gbit/s THz photonic wireless transmission in the 350-GHz band with extended reach. IEEE Photon. Technol. Lett. **30**(11), 1064–1067 (2018)
11. S. Carpenter, D. Nopchinda, M. Abbasi, Z.S. He, M. Bao, T. Eriksson, H. Zirath, A D-Band 48-Gbit/s 64-QAM/QPSK direct-conversion I/Q transceiver chipset. IEEE Trans. Microwave Theory Techniques **64**(4), 1285–1296 (2016)
12. K.K. Tokgoz, S. Maki, J. Pang, N. Nagashima, I. Abdo, S. Kawai, T. Fujimura, Y. Kawano, T. Suzuki, T. Iwai, K. Okada, A. Matsuzawa, A 120Gb/s 16QAM CMOS millimeter-wave wireless transceiver, in *IEEE International Solid—State Circuits Conference—(ISSCC)* pp. 168–170 (2018)

13. P. Rodrìguez-Vìzquez, J. Grzyb, B. Heinemann, U.R. Pfeiffer, A 16-QAM 100-Gb/s 1-M wireless link with an EVM of 17% at 230 GHz in an SiGe technology. IEEE Microwave Wirel. Components Lett. **29**(4), 297–299 (2019)
14. P. RodrìÂguez-Vìzquez, J. Grzyb, B. Heinemann, U.R. Pfeiffer, Performance evaluation of a 32-QAM 1-meter wireless link operating at 220–260 GHz with a data-rate of 90 Gbps, in *2018 Asia-Pacific Microwave Conference (APMC)*, pp. 723–725 (2018)
15. H. Hamada, T. Fujimura, I. Abdo, K. Okada, H.-J. Song, H. Sugiyama, H. Matsuzaki, H. Nosaka, 300-GHz, 100-Gb/s InP-HEMT wireless transceiver using a 300-GHz fundamental mixer, in *IEEE/MTT-S International Microwave Symposium - IMS*, pp. 1480–1483 (2018)
16. F. Boes, T. Messinger, J. Antes, D. Meier, A. Tessmann, A. Inam, I. Kallfass, Ultra-broadband MMIC-based wireless link at 240 GHz enabled by 64GS/s DAC, in *2014 39th International Conference on Infrared, Millimeter, and Terahertz waves (IRMMW-THz)*, pp. 1–2 (2014)
17. I. Ando, M. Tanio, M. Ito, T. Kuwabara, T. Marumoto, K. Kunihiro, Wireless D-band communication up to 60 Gbit/s with 64QAM using GaAs HEMT technology, in *IEEE Radio and Wireless Symposium (RWS)*, pp. 193–195 (2016)
18. I. Kallfass, J. Antes, T. Schneider, F. Kurz, D. Lopez-Diaz, S. Diebold, H. Massler, A. Leuther, A. Tessmann, All active MMIC-based wireless communication at 220 GHz. IEEE Trans. Terahertz Sci. Technol. **1**(2), 477–487 (2011)
19. C. Castro, R. Elschner, J. Machado, T. Merkle, C. Schubert, R. Freund, Ethernet transmission over a 100 Gb/s real-time terahertz wireless link, in *IEEE Globecom Workshops (GC Wkshps)*, pp. 1–5 (2019)
20. Z. Lu, S. Wang, W. Li, S. Jia, L. Zhang, M. Qiao, X. Pang, N. Idrees, M. Saqlain, X. Gao, X. Cao, C. Lin, Q. Wu, X. Yu, 26.8 m 350 GHz wireless transmission of beyond 100 Gbit/s supported by THz photonics, in *2019 Asia Communications and Photonics Conference (ACP)*, pp. 1–3 (2019)
21. X. Li, J. Yu, L. Zhao, W. Zhou, K. Wang, M. Kong, G.-K. Chang, Y. Zhang, X. Pan, X. Xin, 132-Gb/s photonics-aided single-carrier wireless terahertz-wave signal transmission at 450GHz enabled by 64QAM modulation and probabilistic shaping, in *Optical Fiber Communications Conference and Exhibition (OFC)*, pp. 1–3 (2019)
22. I. Dan, G. Ducournau, S. Hisatake, P. Szriftgiser, R.-P. Braun, I. Kallfass, A Terahertz wireless communication link using a Superheterodyne approach. IEEE Trans. Terahertz Sci. Technol. **10**(1), 32–43 (2020)
23. H. Hamada, T. Tsutsumi, H. Matsuzaki, T. Fujimura, I. Abdo, A. Shirane, K. Okada, G. Itami, H.-J. Song, H. Sugiyama, H. Nosaka, 300-GHz-Band 120-Gb/s wireless front-end based on InP-HEMT PAs and mixers. IEEE J. Solid-State Circuits **55**(9), 2316–2335 (2020)
24. I. Dan, P. Szriftgiser, E. Peytavit, J.-F. Lampin, M. Zegaoui, M. Zaknoune, G. Ducournau, I. Kallfass, A 300-GHz wireless link employing a photonic transmitter and an active electronic receiver with a transmission bandwidth of 54 GHz. IEEE Trans. Terahertz Sci. Technol. **10**(3), 271–281 (2020)
25. Y. Feng, B. Zhang, C. Zhi, K. Liu, W. Liu, F. Shen, C. Qiao, J. Zhang, Y. Fan, X. Yang, A 20.8-Gbps dual-carrier wireless communication link in 220-GHz band. China Commun. **18**(5), 210–220 (2021)
26. S. An, Z.S. He, J. Li, X. Bu, H. Zirath, Coded pilot assisted baseband receiver for high data rate millimeter-wave communications. IEEE Trans. Microwave Theory Techniques **68**(11), 4719–4727 (2020)

Chapter 2
Efficient System: Beamspace MIMO-NOMA via D2D Communications

In this chapter, we investigate a full-duplex D2D-aided mmWave MIMO-NOMA broadcasting system. Section 2.1 presents the background and motivation of the research on full-duplex D2D-aided mmWave MIMO-NOMA communication. Section 2.2 introduces a system model of the beamspace MIMO-NOMA for mmWave broadcasting via full-duplex D2D communications. Section 2.3 provides the closed-form expressions of the performance analysis including the outage probability and the ergodic capacity. Section 2.4 validates the theoretical analysis and demonstrates that the system capacity can be improved by the LDM through its numerical results. Section 2.5 offers a conclusion of this chapter.

2.1 Introduction

With the explosive growth of a variety of the commercial broadband and broadcast services, the demands on high data rate, high spectral efficiency, and low latency wireless communications system also increase rapidly. To simultaneously achieve high data throughput, high environmental adaptability, and high physical resource utilization in broadcast network, Advanced Television Systems Committee (ATSC) has standardized the physical layer specification of ATSC 3.0 system for diversified requirements of the market and target devices [1–3]. With the adjustable operating mode, ATSC 3.0 can achieve a goal to balance cellular coverage and system throughput. In addition, the advantages of broadcast and unicast are combined by the cellular network to optimize the system for delivering the broadcast services [4]. The LTE-Advanced Pro has provided the point-to-multipoint transmission by vehicular to everything, Internet of Things and machine-type communication. In 2019, the study on the 5G multicast-broadcast services has been agreed by 3GPP in the future releases of 5G [5].

J. Li et al., *Key Technologies of High Frequency Wireless Communications*,
https://doi.org/10.1007/978-981-96-5894-7_2

The power based LDM is a non-orthogonal technology whose transmitted signal consists of two different layers [6], and the power of each layer is different. In this case, the robust layer can be used to transmit broadcast signals, and the large capactiy layer can be adoped to transmit unicast signal. LDM is a promising technology for the integration of broadband and broadcast [7, 8]. Different from the time division multiplexing (TDM) and frequency division multiplexing (FDM), LDM multiplexes unicast transmission and broadcast transmission in a non-orthogonal fashion to simultaneously achieve high spectrum efficiency and large coverage [9–13]. Considering the diversified quality of service (QoS) requirements of the multiplexed services, LDM supports to choose different transmission power, channel coding and modulation schemes for different layers [14]. In addition, the authors in [15] have proposed a novel low complexity LDM structure for the next generation terrestrial broadcasting systems which can achieve a significant performance gain compared to conventional TDM. Additionally, the N-layered theoretical capacity and the new application scenarios have been introduced in the literature [16]. Furthermore, the performance of LDM with a wide range of stationary channels and the mobile typical urban 6 path (TU-6) channel has been introduced based on the software defined radio platform in the literature [17]. According to the literature [18], the LDM technology has been proposed to improve the LTE evolved multimedia broadcast multicast services (eMBMS). The authors in [19] have shown the performance advantages and potential usage scenarios of combing LDM with multi-radio-frequency channel technologies which can deliver the serves data across two or more radio frequency channels by time slicing and frequency hopping.

D2D communication has been considered as the preferred solution for the content sharing, proximity services and coverage expansion with business purposes in the 3GPP long time evolution (LTE) Release 12 and 13. D2D communication allows a pair of users with closed geographical distances to communicate directly without going though the base station. In the mean time, full-duplex technology is a promising technology to double the spectral efficiency by simultaneous transmission and reception at the same time and the frequency. This feature of full-duplex can be taken advantage at the relay nodes in D2D communication to decrease the time consumption [20]. Since the same time-frequency resources are utilized by the cellular users and D2D users, the power control has been one focus of researchers in recent years. In the literature [21], the centralized and distributed power control themes of the incorporating full-duplex D2D cellular network have been proposed. The self-reference signal of the full-duplex relaying system has been analyzed in [22–24]. And the closed-form approximations of cellular and D2D coverage probabilities have been derived. Meanwhile, the cooperative full-duplex D2D aided cellular network has been introduced in [25] where the cellular user supports D2D communication as a full-duplex relay and an optimal power allocation scheme has been investigated to maximize the system throughout. Moreover, a closed-form expression for the optimal power allocation scheme in the wireless cellular network with full duplex D2D communications has been obtained in the literature [26]. In the literature [27], the bottle-neck effect elimination power algorithm has been proposed to boost the energy efficiency of the full-duplex relay-adided mmWave D2D communications. Furthermore, the literature [28] has proposed a cooperative sub-channel allocation approach in the mmWave networks with D2D communications to improve the resource utilization and network capacity. In addition, the authors in [29] has studied relay-assisted D2D communication in mmWave based 5G networks to balances the trade-off between transmit power and system throughput.

With the increasing demand for a variety of applications and increasingly tight spectrum resources, mmWave has become the optimal choice for high speed, low latency and high throughput communication systems. Due to the large attenuation of the mmWave link, a large scale antenna array is required to improve the antenna gain to ensure the QoS [30]. Furthermore, thanks to the recent development of the antenna circuit design, it is possible to realize large scale antenna arrays including 32 to 256 antennas [31]. The mmWave transceiver architectures are composed of fully digital architecture, analog-only architecture and hybrid analog/digital architecture. In addition, the authors in the literature [32] have proposed a 64-channel massive MIMO transceiver with a fully digital beamforming architecture for mmWave communication which can achieve a downlink data rate of 50.73 Gb/s with 101.5 b/s/Hz. However, because of the high cost and power consumption of high resolution analog to digital converters, it is difficult to realize the fully digital processing at mmWave frequencies with wide bandwidths and large antenna arrays [33–35]. Moreover, analog-only architectures are easy to implement by using the phase shifters. However due to the characteristics of phase shifters, it is difficult to generate multi-beams or support multi-users MIMO communication. Meanwhile, the energy-efficient analog beamforming in mmWave multicast transmission has been proposed in the literature [36] which can realize the optimal beamformer designs and provide asymptotic performance analysis for mmWave multicasting systems. Furthermore, the beamspace MIMO-NOMA for mmWave communications using lens antenna arrays has been investigated in the literature [37], which can achieve a higher spectrum and energy efficiency than the existing system. And the partially-connected hybrid pre-coding design method has been proposed in the literature [38] to maximize the weighted sum of channel gains for mmWave MIMO broadcast channels.

The beamspace MIMO-NOMA and full-duplex D2D communications can be combined to enhance the coverage of broadcast and improve the throughout of unicast in the mmWave broadcasting network. Nonetheless, few researches have analysed the system performance. In this section, we introduce a full-duplex D2D-aided mmWave MIMO-NOMA broadcasting network. Then we analyse the performance of the proposed system under the practical assumptions such as directional beamforming, distanc-dependent pathloss and self-interference of full-duplex communication. The closed-form formulation of the outage probability and the ergodic capacity are derived with Laplace transform technique. Finally, the numerical results indicate that the LDM outperforms the TDM. And the impacts of the antenna array gain and the transmission powers of both base station and the full-duplex link are considered.

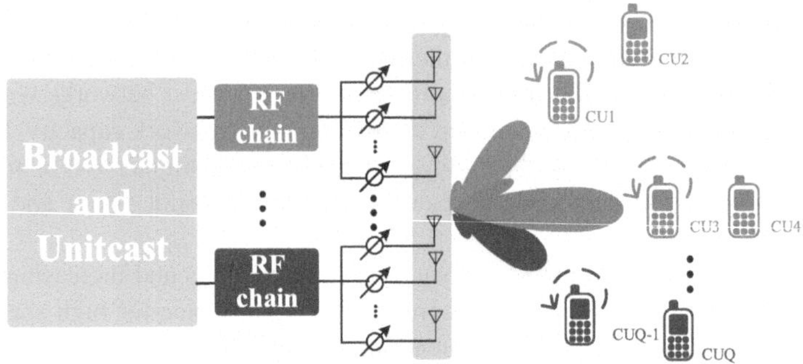

Fig. 2.1 System model of full-duplex D2D-aided mmWave MIMO-NOMA for broadcast and unicast

2.2 System Model

We consider a full-duplex D2D-aided mmWave beamspace MIMO-NOMA system for broadcast and unicast, as shown in Fig. 2.1. The system consists of one base station and Q cellular users which are randomly and uniformly distributed in the cellular network. The base station is equipped with multiple antennas to generate multiple beams with the beamspace methods for cellular users. There are Q cellular user with signal antenna in the cellular network which is denoted as CU_1, CU_2, ..., CU_Q. The cellular users randomly in one beam are paired and communicate with the LDM. In addition, in order to reduce the system complexity, we only consider one beam in the proposed system. There are two users in this beam which are denoted as near user and far user by their distance to the base station. The near user receives the broadcast and unicast signal of the base station, while the far user receiving the broadcast signal from base station and near user. Due to the great attenuation of the millimeter wave system, it is assumed that the near user can provide additional services to the far user through the D2D link with a full-duplex relay.

2.2.1 Layered Division Multiplexing

In order to improve the spectrum efficiency, the two-layer LDM structure was proposed by the ATSC 3.0 to deliver reliable mobile TV services for a large variety of mobile and indoor users. To incorporate LDM in mmWave system, the same mmWave resources are used by the cellular users with the two-layer LDM structure where different layer provides different services. In addition, upper layer (UL) signal is adopted to deliver reliable mobile broadcast services and lower layer (LL) signal is designed to deliver the unicast services. The system diagrams of the transmitter in the two-layer LDM is shown in the Fig. 2.2. The signal for each layer is generated by a separate module, and the signals are superimposed and transmitted in the form

Fig. 2.2 Two-layer LDM transmitter

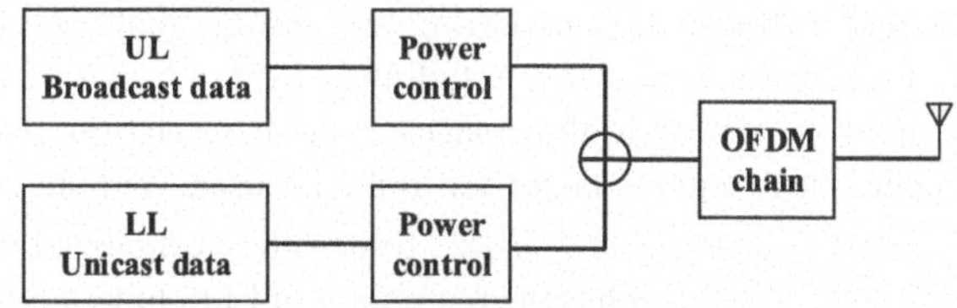

of the OFDM data link. According to the high coverage of the broadcast signal and the high data rate of the unicast signal, high power transmission is adopted by the UL signal, and the LL signal is transmitted with lower power.

In n-th beam, the UL,LL in the n-th beam is denoted as $UL(n)$, $LL(n)$. We assume that $UL(n)$, $LL(n)$ are scheduled on the same radio resource with LDM. We denote x_n as the composite transmission signal of the base station in the n-th beam, which is given by

$$x_n = \sqrt{\lambda_{UL(n)} P_n} S_{UL(n)} + \sqrt{\lambda_{LL(n)} P_n} S_{LL(n)}, \tag{2.1}$$

where $\lambda_{UL(n)}$ and $\lambda_{LL(n)}$ are the power coefficients of UL and LL, respectably, satisfying $\lambda_{UL(n)} + \lambda_{LL(n)} = 1$ and $\lambda_{UL(n)} > \lambda_{LL(n)}$. P_n is the total power in the n-th beam , $S_{UL(n)}$, $S_{LL(n)}$ is the normalized transmission signal of UL and LL in the n-th beam with $\mathbb{E}(\|S_{UL(n)}\|^2) = \mathbb{E}(\|S_{LL(n)}\|^2) = 1$.

2.2.2 Channel Model

The experimental investigations have shown that the mmWave channel is sensitive to the blockage effects. Different channel statistics are proposed by the 3GPP for line of sight (LOS) and non-line of sight (NLOS) links in the simulation. Therefore the rectangle Boolean stochastic blockage model is used to model building blockages by differentiating the LOS and NLOS link [39]. The probability function of a LOS link is denoted as

$$P_L(r) = \exp(-\beta r), \tag{2.2}$$

where r is the distance from transmitter to receiver, β is used to descirbe the size and density of blockages. And the probability function of a NLOS link is $P_N = 1 - \exp(-\beta r)$. Without loss of generality, we assume that the probability function of LOS with different link is independent.

Depending on the Friis equation [33], the mmWave can not propagate well in free space. The large scale fading $L(r)$ in dB of the mmWave link ,which is introduced in [40], is modeled as

$$L(r) = 32.4 + 20 \log(f_c) + 10\alpha \log(r), \tag{2.3}$$

where f_c is the carrier frequency, $\alpha \in \{\alpha_L, \alpha_N\}$ is the path loss exponents of LOS, NLOS, and r is the distance between the transmitter and receiver.

As for the small scale fading, we assume that each link between the transmitter and receiver is Nakagami-m fading [41]. $h_{u_1}, h_{u_2}, h_{I_1}$ and h_D are denoted as the link of the base station to the near user, the base station to the far user, the near user to the near user and the near user to the far user. Without loss of generality, we assume that $\|h_{u_1}\|^2, \|h_{u_2}\|^2$ and $\|h_D\|^2$ follow the normalized Gamma random varible with $N \in \{N_L, N_N\}$ which is the parameter of LOS link or NLOS link. Then the h_{I_1} is free of fading. And the probability density function $f(x)$ and the cumulative distribution function $F(x)$ of $\|h_{u_1}\|^2$ are respectively given by

$$f(x) = \frac{N^N}{\Gamma(N)} x^{N-1} e^{-Nx}, \tag{2.4}$$

and

$$F(x) = \begin{cases} 1 - \sum_{J=0}^{N-1} \frac{N^j}{j!} x^j e^{-Nx}, & x \geq 0 \\ 0, & \text{otherwise} \end{cases}. \tag{2.5}$$

2.2.3 Large Antenna Arrays

Due to the large attenuation of the mmWave link, the high gain and high directional beam is generated by the base station to ensure high QoS. To simplify the mathematical analysis, we assume the actual antenna model as the sectorized antenna model which means the array gain within the half-power beam width (main lobe gain) is replaced by the maximum power gain and the other direction of arrivals (side lobe gain) is replaced by the first minor maximum gain. Then the total array gain $G(n)$ is denoted as [42],

$$G(n) = \begin{cases} \Psi, & \text{main lobe} \\ \psi, & \text{side lobe} \end{cases}, \tag{2.6}$$

where $\Psi = M$, $\psi = 1/sin^2(\frac{3\pi}{2\sqrt{M}})$ and M is the number of antennas.

2.2.4 Receive Signal

For the broadcast and unicast transmissions, not only receiving the signal from the base station, the near user but also receives the self-interference signal by itself due to the full duplex. Meanwhile, the far user receives the broadcast data from the base station and the near user. Without loss of generality, the near user and far user in one beam are introduced as followed. Then the received signal of near user y_{u1} and the received signal of far user y_{u2} can be respectively formulated as

$$y_{u1} = \frac{h_{u1}G(n)x_n}{L(d_1)} + \frac{h_{I_1}\sqrt{P_D}S_{UL}}{L_{I_1}} + n_{u1}, \tag{2.7}$$

and

$$y_{u2} = \frac{h_{u2}G(n)x_n}{L(d_2)} + \frac{h_D\sqrt{P_D}S_{UL}}{L(d_2 - d_1)} + n_{u2}, \qquad (2.8)$$

where h_{u1}, h_{u2} is the channel gain from the base station to the near user and the far user respectively, $\|h_{u1}\|^2 > \|h_{u2}\|^2$. h_{I_1} is the channel gain from the near user to the near user. h_D is the channel gain from the near user to the far user. x_n is the superimposed signal by the broadcast and unicast data in the n-th beam. d_1, d_2 are the distance from base station to the near user and far user. In addition, the distance between the two users is approximated by the difference between their distance to the base station due to the narrow beam width of mmWave. L_{I_1} is the self-interference signal link attenuation of near user. P_D is the transmitted power from near user to far user. S_{UL} is the broadcast signal from the near user to the far user, $\mathbb{E}(\|S_{UL(n)}\|^2) = 1$. In addition, n_{u1}, n_{u2} are the independent and identically distributed normalized white Gaussian noise which are denoted as $n_{u1}, n_{u2} \sim CN(0, 1)$.

2.3 Performance Analysis

In this section, the performance analysis of full-duplex D2D-aided mmWave beamspace MIMO-NOMA is proposed. The outage probability of users and the ergodic capacities of broadcast and unicast services are introduced as followed. In the proposed system, firstly, the base station transmits the LDM signal to the cellular users. Next, the near user receives the broadcast and unicast signal with the perfect successive interference cancellation method. Then the full-duplex mode with D2D communication is adopted to transmit broadcast signals to far user. Finally, the far user receives the signals from the base station and the D2D links. Two events are considered in the system as followed.

Event 1: The received signal of the near user is described as event one. The near user receives the signal sent by the base station and the self-interference signal of the near user. The broadcast signal is first solved by regarding the unicast signal and the self-interference signal as interference. Then the near user cancels the broadcast signal and resolves the unicast signal in the presence of self-interference signals. The near user first decodes the broadcast data and remove it with γ_{B1} and then the near user obtains the unicast data with γ_{U1}.

$$\gamma_{B1} = \frac{\lambda_{UL(n)}G(n)P_n/L(d_1)\|h_{u1}\|^2}{\lambda_{LL(n)}G(n)P_n/L(d_1)\|h_{u1}\|^2 + P_D/L_{I_1}\|h_{I_1}\|^2 + \sigma_1^2}, \qquad (2.9)$$

$$\gamma_{U1} = \frac{\lambda_{LL(n)}G(n)P_n/L(d_1)\|h_{u1}\|^2}{P_D/L_{I_1}\|h_{I_1}\|^2 + \sigma_1^2}. \qquad (2.10)$$

Event 2: The received signal of the far user is described as event two. The far user receives the broadcast data from the base station and the near user at the same time. We assume that these two signals are solvable at far user, so that the maximal ratio combing method can be adopted. Therefore, the far user can acquire the broadcast data with γ_{B2}.

$$\gamma_{B2} = \frac{\lambda_{UL(n)} G(n) P_n/L(d_2)\,\|h_{u2}\|^2 + P_D/L(d_2 - d_1)\,\|h_{I_D}\|^2}{\lambda_{LL(n)} G(n) P_n/L(d_2)\,\|h_{u2}\|^2 + \sigma_2^2}. \tag{2.11}$$

2.3.1 Outage Probability

The outage probability of near user and far user is considered in this part. The target rate of the broadcast of near user, unicast of near user and broadcast of far user are denoted as R_1, R_2 and R_3, respectively. And the outage probability of near user and far user are denoted as P_1 and P_2 as follows,

$$\begin{aligned} P_1 &= P(\log(1 + \gamma_{B1}) < R_1 \;\text{or}\; \log(1 + \gamma_{U1}) < R_2) \\ &= 1 - P(\log(1 + \gamma_{B1}) \geq R_1)P(\log(1 + \gamma_{U1}) \geq R_2), \end{aligned} \tag{2.12}$$

$$P_2 = P(\log(1 + \gamma_{B2}) < R_3). \tag{2.13}$$

First of all, we consider the term $P(\log(1 + \gamma_{B1}) < R_1)$. By denoting $\epsilon_1 = 2^{R_1} - 1$, this term can be formulated as

$$\begin{aligned} &P(\log(1 + \gamma_{B1}) < R_1) \\ &= P(\frac{\lambda_{UL(n)} P_n G(n)/L(d_1)\,\|h_{u1}\|^2}{\frac{\lambda_{LL(n)} P_n G(n)}{L(d_1)}\,\|h_{u1}\|^2 + P_D/L_{I_1}\,\|h_{I_1}\|^2 + \sigma_1^2} < \epsilon_1). \end{aligned} \tag{2.14}$$

Then we denote $\eta = \frac{\epsilon_1 L(d_1) N_m}{(\lambda_{UL(n)} - \lambda_{LL(n)}\epsilon_1) P_n G(n)}$, $I_1 = \frac{P_D \|h_{I_1}\|^2}{L_{I_1}}$, $\|h_{u1}\|^2 \sim Gamma(N_m)$, $N_m \in \{N_L, N_N\}$. When the $\eta \leq 0$, the (2.14) is equal to 1. Otherwise, the (2.14) can be simplified to

$$\begin{aligned} &P(\|h_{u1}\|^2 < \frac{\eta}{N_m}(I_1 + \sigma_1^2)) \\ &= \mathbb{E}_{d_1}\{P(\|h_{u1}\|^2 < \frac{\eta}{N_m}(I_1 + \sigma_1^2))|d_1\} \\ &= \mathbb{E}_{d_1}\{\sum_{m \in \{L,N\}} P_m(d_1)P(\|h_{u1}\|^2 < \frac{\eta}{N_m}(I_1 + \sigma_1^2)|d_1, m)\}. \end{aligned} \tag{2.15}$$

Since the probability of LOS link is related to the distance d_1, the conditional probability with a fixed d_1 and m will be firstly solved as,

$$P(\|h_{u1}\|^2 < \frac{\eta}{N_m}(I_1 + \sigma_1^2)|d_1, m)$$

$$\overset{(a)}{=} 1 - \sum_{J=0}^{N-1} \frac{1}{j!}(\eta(I_1 + \sigma_1^2)^j)e^{-\eta(I_1+\sigma_1^2)} \tag{2.16}$$

$$\overset{(b)}{=} 1 - \sum_{J=0}^{N-1} \frac{(-\eta)^j}{j!}(e^{-\eta(I_1+\sigma_1^2)})^{(j)},$$

where step-(a) is due to the cumulative distribution function of $\|h_{u1}\|^2$. And the j in the upper right corner of the step-(b) represents the j-th derivative of the formula.

Because the users are randomly and uniformly distributed in the cellular of radius R and satisfy $d_1 < d_2$, the pdf of d_1 and d_2 can be given by $f_{d_1}(r) = \frac{4}{R^2}r - \frac{4}{R^4}r^3$ and $f_{d_2}(r) = \frac{4}{R^4}r^3$, respectively, using order statistics. We set the $L(\eta) = e^{-\eta(I_1+\sigma_1^2)}$. By substituting (2.16) into (2.15), the item $P(\log(1 + \gamma_{B1}) < R_1)$ can be denoted as

$$P(\log(1 + \gamma_{B1}) < R_1)$$

$$= \mathbb{E}_{d_1}\{\sum_{m\in\{L,N\}} P_m(d_1)(1 - \sum_{J=0}^{N-1} \frac{(-\eta)^j}{j!}(L(\eta))^{(j)})\} \tag{2.17}$$

$$= \sum_{m\in\{L,N\}} \sum_{J=0}^{N-1} \int_0^R (1 - P_m(r))\frac{(-\eta)^j}{j!}(L(\eta))^{(j)}(\frac{4}{R^2}r - \frac{4}{R^4}r^3))\mathrm{d}r$$

The Gaussian-Chebyshev approximation can be adopted to calculate the integral in (2.17). And we set $\theta_i = \cos(\frac{(2i-1)/pi}{2M})$, $r_i = \frac{R}{2}(\theta_i + 1)$, where M is the parameter of the Gaussian-Chebyshev approximation. Finally we can get the final expression,

$$P(\log(1 + \gamma_{B1}) < R_1)$$

$$= \sum_{m\in\{L,N\}} \sum_{j=0}^{N-1} \sum_{i=1}^{M} \frac{R\pi\sqrt{1 - \theta_i^2}}{2M}(1 - P_m(r_i))\frac{(-\eta)^j}{j!}(L(\eta))^{(j)} f_{d_1}(r_i)). \tag{2.18}$$

Similarly, the item $P(\log(1 + \gamma_{U1}) < R_1)$ can also be derived using (2.18) by simply setting $\eta = \frac{\epsilon_2 L(d_1)N_m}{(\lambda_{LL(n)}\epsilon_1)P_n G(n)}$, $\epsilon_2 = 2^{R_2} - 1$. Finally, P_1 can be derived by (2.12).

Next, we consider the outage probability of far user P_2. We set $\epsilon_3 = 2^{R_3} - 1$, then the outage probability P_2 can be formulated as

$$P_2 = P(\frac{\lambda_{UL(n)}G(n)P_n/L(d_2)\|h_{u2}\|^2}{\lambda_{LL(n)}G(n)P_n/L(d_2)\|h_{u2}\|^2 + \sigma_2^2}$$

$$+ \frac{P_D/L(d_2 - d_1)\|h_{I_D}\|^2}{\lambda_{LL(n)}G(n)P_n/L(d_2)\|h_{u2}\|^2 + \sigma_2^2} < \epsilon_3) \tag{2.19}$$

$$= P(\|h_{I_D}\|^2 < \frac{\beta}{N_m}(I_2 - \xi)),$$

where

$$\begin{cases} \beta = (\epsilon_3 \lambda_{LL(n)} - \lambda_{UL(n)}) \frac{G(n) P_n L(d_2 - d_1)}{L(d2) P_D} \\ \xi = \frac{L(d_2) \epsilon_3 \sigma_2^2}{(\lambda_{UL(n)} - \epsilon_3 \lambda_{LL(n)}) G(n) P_n} \\ I_2 = \|h_{u2}\|^2 \end{cases} \tag{2.20}$$

The outage probability of far user can be calculated as

$$\begin{aligned} P_2 &= \mathbb{E}_{d_1,d_2,I_2}\{ \sum_{m \in \{L,N\}} P_m(d_2) P(\|h_{I_D}\|^2 < \frac{\beta}{N_m}(I_2 - \xi)|d_1, d_2, m)\} \\ &= \sum_{m \in \{L,N\}} \sum_{j=0}^{N-1} \mathbb{E}_{d_1,d_2,I_2}\{1 - P_m(d_2) \frac{(-\beta)^j}{j!} (e^{-\beta(I_2 - \xi)})^{(j)}\}. \end{aligned} \tag{2.21}$$

With the different values of system parameters, β can be divided into two cases to discuss. When $\beta > 0$, the $\epsilon \lambda_{LL} - \lambda_{UL} > 0$, the $P_1 = 1$. It means that the near user can not receive the broadcast signals all the time. In our actual deployment, we need to satisfy the formula $\epsilon \lambda_{LL} - \lambda_{UL} < 0$ between the power allocation ratio and the broadcast target rate to communicate properly.

When $\beta < 0$, if $I_2 \geq \xi$, $P_2 = 0$. And if $I_2 < \xi$, the P_2 can be formulated as,

$$\begin{aligned} \mathbb{E}_{I_2}\{e^{-\beta I_2}\} &= \int_0^{\xi} e^{-\beta x} \frac{N_m^{N_m}}{\Gamma(N_m)} x^{N_m - 1} e^{-N_m x} dx \\ &= \frac{N_m^{N_m}}{\Gamma(N_m)} e^{-(N+\beta)\xi} \left(\sum_{k=0}^{N-1} \frac{(-1)k!\binom{N-1}{k}}{(N+\beta)^{k+1}} \xi^{N-1-k} \right) + (1 + \frac{\beta}{N_m})^{-N_m} \end{aligned} \tag{2.22}$$

Next, the pdf of d_1 and d_2 can be used to calculated the expectation of d_1 and d_2, which is denoted as $f_{d_1,d_2}(r_1.r_2) = \frac{8 r_1 r_2}{R^4}$. We set the $L(\beta, \xi) = e^{\beta \xi} \mathbb{E}_{I_2}\{e^{-\beta I_2}\}$. We can get

$$\begin{aligned} P_2 &= \sum_{m \in \{L,N\}} \sum_{j=0}^{N-1} \mathbb{E}_{d_1,d_2}\{1 - P_m(d_2) \frac{(-\beta)^j}{j!} (L(\beta, \xi))^{(j)}\} \\ &= \sum_{m \in \{L,N\}} \sum_{j=0}^{N-1} \int_0^R \int_0^{r_2} (1 - P_m(r_2) \frac{(-\beta)^j}{j!} \end{aligned} \tag{2.23}$$

$$(L(\beta, \xi))^{(j)} f_{d_1,d_2}(r_1, r_2)) dr_1 dr_2$$

Finally, we use Gaussian-Chebyshev approximation to calculate the integral (2.23). We define the parameters related to this approximation as follows:

$$\begin{cases} \theta_{1i} = \cos(\frac{(2i-1)\pi}{2M}) \\ r_{1i} = \frac{r_2}{2}(\theta_{1i} + 1) \\ \theta_{2k} = \cos(\frac{(2k-1)\pi}{2M}) \\ r_{2k} = \frac{R}{2}(\theta_{2k} + 1) \end{cases} \tag{2.24}$$

where M is the parameter of the Gaussian-Chebyshev approximation. The colosed-form expression of P_2 can be finally given by,

$$P_2 = \sum_{m \in \{L, N\}} \sum_{j=0}^{N-1} \sum_{i=1}^{M} \sum_{k=1}^{M} \frac{R\pi}{2M} \sqrt{1 - \theta_{2k}^2} \frac{r_{2k}\pi}{2M} \sqrt{1 - \theta_{1i}^2}$$

$$(1 - P_m(r_{2k})) \frac{(-\beta_{i,k})^j}{j!} (L_{i,k}(\beta, \xi))^{(j)}) f_{d_1, d_2}(r_{1i}, r_{2k}) \tag{2.25}$$

2.3.2 Ergodic Capacity

In the following, we analyse the ergodic capacity of two users which represents the ability of the entire system to deliver information. The ergodic capacity of the system, denoted as C_{Erg}, can be obtained by Shannon's Law,

$$C_{\text{Erg}} = \mathbb{E}[\log_2(1 + \gamma_{U_1})] + \mathbb{E}[\log_2(1 + \gamma_{B_2})]$$

$$= \mathbb{E}[\log_2(1 + \frac{\lambda_{LL(n)} G(n) P_n / L(d_1) \|h_{u1}\|^2}{P_D / L_{I_1} \|h_{I_1}\|^2 + \sigma_1^2}] + \mathbb{E}[\log_2(1 +$$

$$\frac{\lambda_{UL(n)} G(n) P_n / L(d_2) \|h_{u2}\|^2 + P_D / L(d_2 - d_1) \|h_{I_D}\|^2}{\lambda_{LL(n)} G(n) P_n / L(d_2) \|h_{u2}\|^2 + \sigma_2^2})] \tag{2.26}$$

According to the literature [43], the capacity formula can be approximated by.

$$\mathbb{E}[\ln(1 + x)] \approx \ln(1 + \mathbb{E}[x]) - \frac{\mathbb{E}[x^2] - (\mathbb{E}[x])^2}{2(1 + \mathbb{E}[x])^2}, \tag{2.27}$$

which reveals that the ergodic capacity can be approximated by a composite function of $\mathbb{E}[x]$ and $\mathbb{E}[x^2]$. Therefore, the following part focus on the derivation of $\mathbb{E}[x]$ and $\mathbb{E}[x^2]$.

Then we solve numerator part of the second item in Eq. (2.26).

$$\mathbb{E}_{h_{u2}, h_{I_D}, d1, d2}[x] = \mathbb{E}_{h_{u2}, h_{I_D}, d1, d2}[G(n) P_n / (L(d_2) \sigma_2^2)$$

$$\|h_{u2}\|^2 + P_D / (L(d_2 - d_1) \sigma_2^2) \|h_{I_D}\|^2]$$

$$= \mathbb{E}_{d1, d2}[\sum_{u_2 \in \{L, N\}} P_{u_2}(d_2)(G(n) \frac{P_n}{L(d_2) \sigma_2^2} + \frac{P_D}{L(d_2 - d_1) \sigma_2^2})] \tag{2.28}$$

$$= \sum_{u_2 \in \{L, N\}} \int_0^R \int_0^{r_2} P_{u_2}(r_2) L(r_1, r_2) f_{d_1, d_2}(r_1, r_2) dr_1 dr_2,$$

where $a = G(n)P_n/\sigma_2^2$ and $b = P_D/\sigma_2^2$. The term $L(r_1, r_2) = \frac{a}{r_2^{\alpha_{u2}}} + \frac{b}{(r_2-r_1)^{\alpha_{I_D}}}$. The Gaussian-Chebyshev approximation can be also adopted to calculate the integral in (2.28). By setting (2.24), where M is the parameter of the Gaussian-Chebyshev approximation, we can derive $\mathbb{E}[x]$ as followed,

$$\mathbb{E}[x] = \sum_{u_2 \in \{L,N\}} \sum_{m=1}^{M} \sum_{i=1}^{M} (\frac{\pi}{2M})^2 \sqrt{1 - \theta_i^2} R \sqrt{1 - \theta_m^2} r_m \tag{2.29}$$

$$P_{u_2}(r_m) L(r_i, r_m) f_{d_1,d_2}(r_{1i}, r_{2m}).$$

Next, $\mathbb{E}[x^2]$ is calculated by firstly solving the expectation of $\|h_{u2}\|^2$ and $\|h_{I_D}\|^2$, and then solving the expectation with respect to d_1, d_2 as follows,

$$\mathbb{E}_{h_{u2},h_{I_D},d1,d2}[x^2] = \mathbb{E}_{h_{u2},h_{I_D},d1,d2}[(G(n)P_n/(L(d_2)\sigma_2^2)$$

$$\|h_{u2}\|^2 + P_D/(L(d_2 - d_1)\sigma_2^2) \|h_{I_D}\|^2)^2]$$

$$= \sum_{u_2 \in \{L,N\}} \int_0^R \int_0^{r_2} P_{u_2}(r_2) L(r_1, r_2) f_{d_1,d_2}(r_1, r_2) \mathrm{d}r_1 \mathrm{d}r_2, \tag{2.30}$$

where $c = G(n)P_n/\sigma_2^2$, $d = P_D/\sigma_2^2 0$. $L(r_1, r_2) = (\frac{c}{r_2^{\alpha_{u2}}})^2 (N_{u2} + 1)/N_{u2} + (\frac{d}{(r_2-r_1)^{\alpha_{I_D}}})^2 (N_{I_D} + 1)/N_{I_D} + 2 \cdot \frac{c}{r_2^{\alpha_{u2}}} \cdot \frac{d}{(r_2-r_1)^{\alpha_{I_D}}}$. By setting (2.24), where M is the parameter of the Gaussian-Chebyshev approximation, we can derive $\mathbb{E}[x]$ as followed,

$$\mathbb{E}[x^2] = \sum_{u_2 \in \{L,N\}} \sum_{m=1}^{M} \sum_{i=1}^{M} (\frac{\pi}{2M})^2 \sqrt{1 - \theta_i^2} R \sqrt{1 - \theta_m^2} r_m \tag{2.31}$$

$$P_{u_2}(r_m) L(r_i, r_m) f_{d_1,d_2}(r_{1i}, r_{2m}).$$

By substituting (2.29) and (2.31) into (2.27), the numerator part of the second item in equation (2.26) can be calculated. The first item in equation (2.26) is

$$\mathbb{E}[\log_2(1 + \frac{\lambda_{LL(n)} G(n) P_n / L(d_1) \|h_{u1}\|^2}{P_D/L_{I_1} \|h_{I_1}\|^2 + \sigma_1^2})]. \tag{2.32}$$

We set $a = \frac{\lambda_{LL(n)} G(n) P_n}{P_D/L_{I_1} \|h_{I_1}\|^2 + \sigma_1^2}$. Similar to the (2.29) and (2.31), the $\mathbb{E}[x]$ and $\mathbb{E}[x^2]$ should be calculated first. Then we can get

$$\mathbb{E}[x] = \sum_{u_2 \in \{L,N\}} \sum_{i=1}^{M} (\frac{\pi}{2M}) R \sqrt{1 - \theta_i^2} P_{u_1}(r_i) L(r_i) f_{d_1}(r_i), \tag{2.33}$$

where $L(r_i) = 1/(r)^{\alpha_{u_1}}$.

$$\mathbb{E}[x^2] = \sum_{u_2 \in \{L,N\}} \sum_{i=1}^{M} (\frac{\pi}{2M}) R \sqrt{1 - \theta_i^2} P_{u_1}(r_i) L(r_i) f_{d_1}(r_i), \qquad (2.34)$$

where $L(r_i) = (N_{u1} + 1)/(N_{u1}(r)^{2\alpha_{u_1}})$.

Then we can submit (2.33) and (2.34) into (2.27) to calculate the first item in equation (2.26). As for the Denominator of the second item in equation (2.26), we only change the value of a of (2.33) and (2.34). Finally, we conmbina all the item in the equation (2.26), and the ergodic capacity can be obtained in a closed-form.

2.4 Numerical Results

In this section, the outage probability and the ergodic capacity of the beamspace MIMO-NOMA millimeter wave broadcasting network via full-duplex D2D communications are investigated. The impact of the transmission power of base station, the antenna number of base station, the transmission power of near user and the power allocation on the outage probability and ergodic capacity are introduced as followed. The traditional TDM scheme without D2D communications and the LDM without D2D communications are used for comparison to analyze the performance of the proposed system. In particular, the time slot is equally divided by the two users. Hence the capacity of this scheme is R_{TDM}, which is denoted as

$$R_{TDM} = \frac{1}{2}(\log(1 + \gamma_1) + \log(1 + \gamma_2)). \qquad (2.35)$$

The numerical results of the proposed system performance are considered with Monte Carlo simulations. The simulation parameters and setting of the proposed system are shown at Table 2.1.

2.4.1 Outage Probability

The outage probability of the near user and the far user are analysed in this section.

In Fig. 2.3, the impacts of the base station transmission power on the outage probability are analysed. It can be seen from the figure that the Monte Carlo simulation is consistent with the proposed closed-formed expression. With the increasing of the transmission power of base station, the outage probability of near user and far user are decreasing, where the LDM outperforms the traditional TDM mode. In addition, compared with LDM without D2D, the outage probability of the near user increases slightly, but the outage probability of the far user decreases by 30 dB. Hence, increasing the transmission power of base station can be considered to reduce the outage

Table 2.1 The simulation parameters and setting of the proposed system

Parameter	Value
Carrier frequency	28 GHz
Bandwidth	200 MHz
Coverage radius	30
The number of base station antenna	64
The transmission power of the near user	5 dbm
The isolation of the transmitted and received in the near user	125 dB
The power coefficient of the LDM	0.15
The small scale fading N_L	2
The small scale fading N_N	3
The pass loss exponents α_L	2
The pass loss exponents α_N	3
The blockage parameter	0.001
The target rate R_1	3.8 bit/s/Hz
The target rate R_2	10 bit/s/Hz
The target rate R_3	3.8 bit/s/Hz

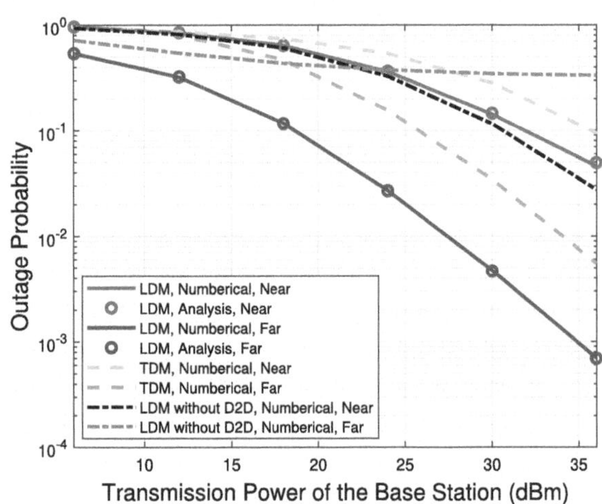

Fig. 2.3 Impact of the base station transimission power on outage probability

probability further to enlarge the coverage and enhance the throughput of the whole system.

In Fig. 2.4, due to the self-interference of the near user, the outage probability of the near user slightly increase with the transmission power of the near user increasing. In addition, the outage probability of far user keeps a downward tendency which benefits from the improvement of the signal-to-noise ratio (SNR) at the receiving end of the D2D communication. The SNR went from -10 dB to 15 dB and the outage probability of far user dropped by about half. In the practical application,

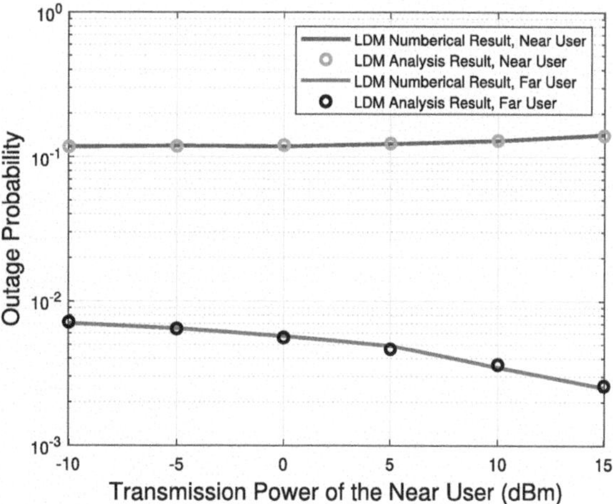

Fig. 2.4 Impact of the near user transimission power on outage probability

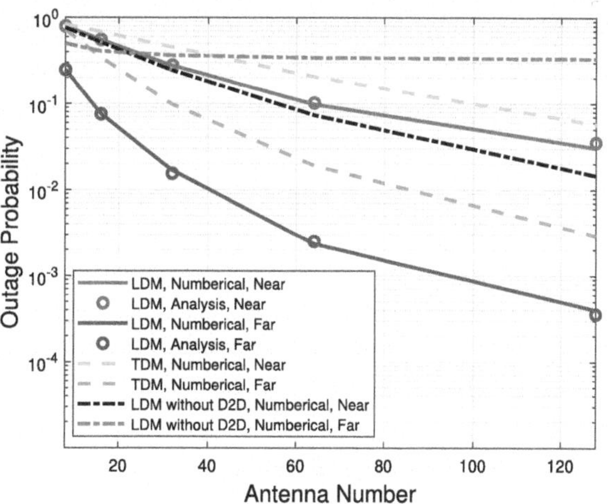

Fig. 2.5 Impact of the antenna number on outage probability

the transmission power of the near user can be increased with a small impact on the near user to reduce the outage probability of the far user and further to enhance the cellular coverage.

In Fig. 2.5, the effect of the antenna number of base station on the outage probability is considered. The simulation results verify that the large scale antenna arrays have a obvious affection on the outage probability at the mmWave MIMO-NOMA broadcasting network. The outage probability of the proposed LDM is 10 dB lower than the TDM. In actual system deployment, a large scale antenna arrays is adopted to compensate for the problem of large attenuation of millimeter wave in signal transmission, which can improve the transmission rate of the system, reduce delay and ameliorate the QoS.

In Fig. 2.6, it has shown that power allocation has a greater impact on near user, because the lower layer is used to transmit unicast signals of the near user. When the

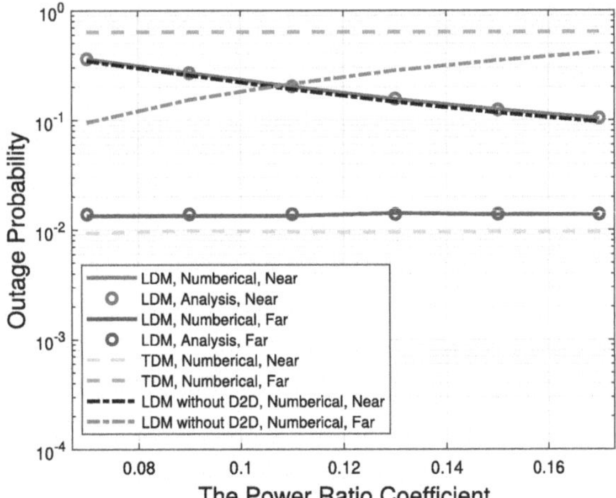

Fig. 2.6 Impact of power ratio coefficient on outage probability

power ratio is small, the outage probability increases. Moreover, far user not only receives signals from the base station, but also receives signals from the D2D link, so the impact is small. In order to reduce the outage probability of near user, the power ratio coefficient can be appropriately increased.

Simulation results show that the performance of LDM is better than the TDM. It can be seen that the non-orthogonal multiple access method can achieve high spectrum efficiency and high throughput wireless transmission services by slightly increasing the complexity of the receiver while the hardware transmission structure is unchanged. In addition, it can be seen that the transmission power and the number of antennas of the base station can be increased to reduce the outage probability.

2.4.2 Ergodic Capacity

In this section, the ergodic capacity of the proposed system which is the sum rate of the near user and the far user is considered. Four factors such as the transmission power of the base station, the transmission power of the near user, the number of antenna and the power allocation are analysed as followed.

In Fig. 2.7, the impact of base station transmission power on the ergodic capacity is investigated. It can be seen that the simulation results match to the closed form expression of the proposed LDM which is better than the TDM and the LDM without D2D communications. In addititon, the ergodic capacity is approximately linearly related to the transmission power of the base station. The ergodic capacity of the proposed LDM is about 1.2 bit/s/Hz higher than the TDM. In the actual deployment, increasing the transmission power of the base station as much as possible to enlarge the coverage and improve the throughput of the entire system under the actual constraints such as human safety, power consumption and heat dissipation, etc.

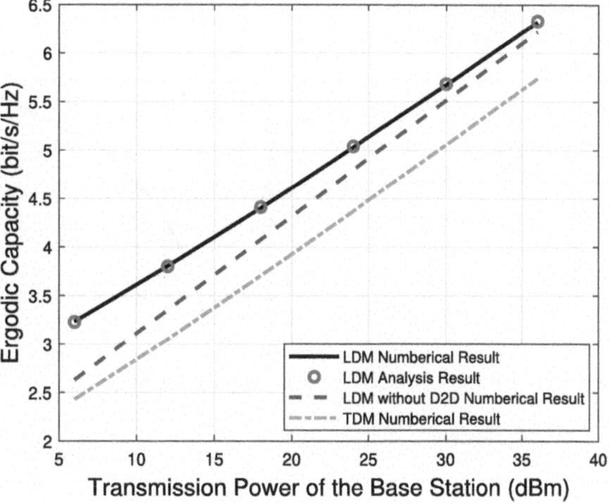

Fig. 2.7 Impact of the base station transimission power on ergodic capacity

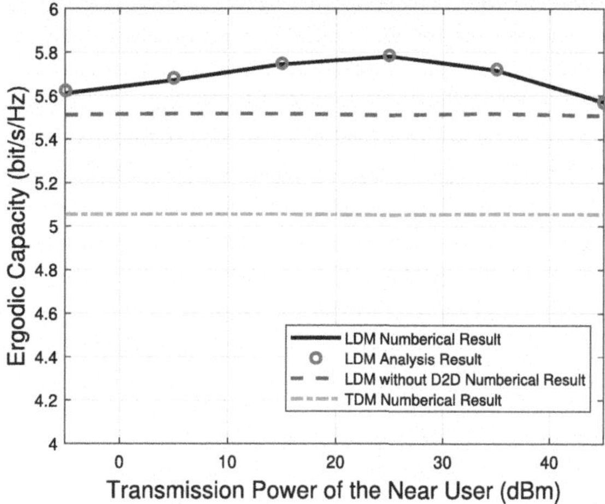

Fig. 2.8 Impact of the near user transimission power on ergodic capacity

In Fig. 2.8, it is shown that as the transmission power of the near user increasing, the ergodic capacity increases first, and decrease next. When the transmission power of the near user is small, the self-interference of the near user has little effect on the system ergodic capacity. At this time, the decisive factor for the total system ergodic capacity is the capacity of the far user. The capacity of the far user increases with the transmission power of the near user increasing. However, when the transmission power of the near user is large, the self-interference becomes an important factor affecting the system capacity. Therefore, in actual deployment, multiple aspects of the entire system must be fully considered to determine the transmission power of the near user to ensure the maximum system ergodic capacity. First, the parameters of the cellular network can be fixed. Then we can use the power of full-duplex transmission

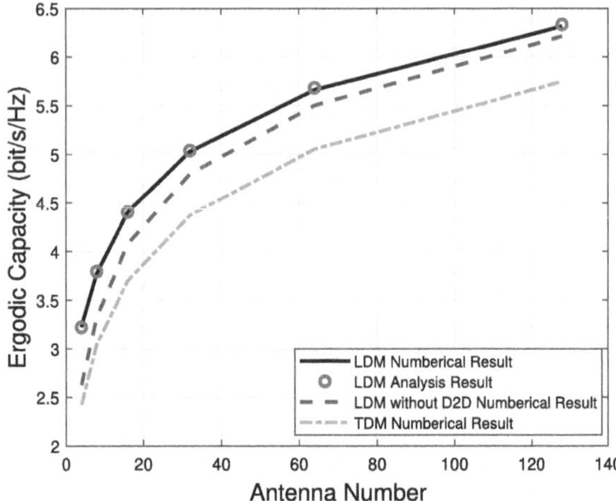

Fig. 2.9 Impact of the antenna number on ergodic capacity

link as a variable to derive the system capacity of the cellular network, and make the derivative equal to zero. Finally, the optimal transmission power can be adopted.

In Fig. 2.9, the system ergodic capacity increases in logarithmic form as the number of antennas increases. Meanwhile, the simulation results indicates that the ergodic capacity of LDM is improved better than the TDM. The proposed LDM is 0.2 bit/s/Hz higher than the LDM without D2D communications when the antenna number larger than 64. The millimeter wave is easy to integrate large scale antenna arrays thanks to its short wavelength. In addition, we need to make a trade-off between the increments of the system capacity and the power consumption of the large scale antennas and radio frequency chains.

In Fig. 2.10, when the power factor changes, the ergodic capacity remains almost unchanged. Meanwhile, the simulation results indicates that the ergodic capacity of

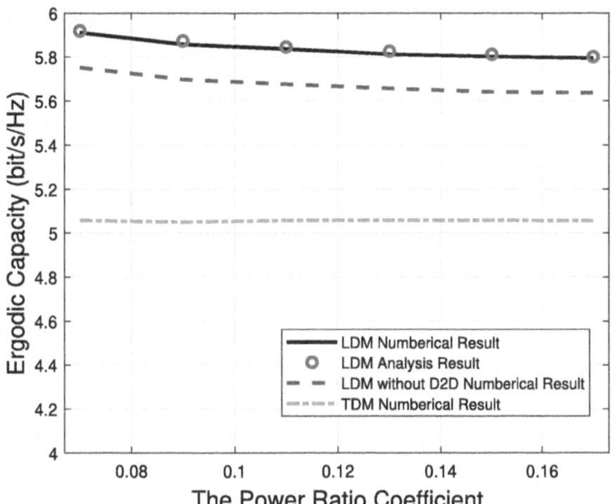

Fig. 2.10 Impact of power ratio coefficient on ergodic capacity

the proposed LDM is improved better than the TDM and the LDM without D2D communications.

The impact factors of the system ergodic capacity are considered. The system ergodic capacity will increase with the transmission power of base station and the antenna number of base station increasing. Furthermore, the transmission power of the near user should be optimized to enlarge the system ergodic capacity.

2.5 Conclusion

In this section, the outage probability and the ergodic capacity of the mmWave MIMO-NOMA broadcasting network with full-duplex D2D communications are presented. The accuracy of the analysis is then verified by detailed numerical results. Compared with the conventional TDM-based and LDM without D2D scheme, our proposed scheme achieves significant performance gains with respect to both outage probability and the ergodic capacity. We also observe that the system performance can be improved with higher transmission power and the larger antenna array at the base station. Besides, the transmission power of the full-duplex communication should be optimized to ensure low self-interference and large capacity.

References

1. L. Fay, L. Michael, D. Gómez-Barquero, N. Ammar, M.W. Caldwell, An overview of the atsc 3.0 physical layer specification. IEEE Trans. Broadcasting **62**, 159–171 (2016)
2. Y. Wu, B. Rong, K. Salehian, G. Gagnon, Cloud transmission: a new spectrum-reuse friendly digital terrestrial broadcasting transmission system. IEEE Trans. Broadcasting **58**, 329–337 (2012)
3. B. Rong, Y. Qian, K. Lu, H. Chen, M. Guizani, Call admission control optimization in wimax networks. IEEE Trans. Veh. Technol. **57**, 2509–2522 (2008)
4. F. Hartung, U. Horn, J. Huschke, M. Kampmann, T. Lohmar, M. Lundevall, Delivery of broadcast services in 3g networks. IEEE Trans. Broadcasting **53**, 188–199 (2007)
5. D. Gomez-Barquero, D. Navratil, S. Appleby, and M. Stagg, Point-to-multipoint communication enablers for the fifth generation of wireless systems. IEEE Commun. Standards Mag. **2**, 53–59 (2018)
6. N. Ye, X. Li, H. Yu, L. Zhao, W. Liu, X. Hou, Deepnoma: a unified framework for noma using deep multi-task learning. IEEE Trans. Wirel. Commun., 1 (2020)
7. J. Zhao, D. Gündüz, O. Simeone, D. Gómez-Barquero, Non-orthogonal unicast and broadcast transmission via joint beamforming and ldm in cellular networks. IEEE Trans. Broadcasting, 1–13 (2019)
8. N. Ye, H. Han, L. Zhao, A.-h. Wang, Uplink nonorthogonal multiple access technologies toward 5g: a survey. Wirel. Commun. Mobile Comput. (2018)
9. B. Xiao, K. Xiao, Z. Chen, B. Xia, H. Liu, Joint design for modulation and constellation labels in multiuser superposition transmission. IEEE Trans. Broadcasting **65**, 245–259 (2019)
10. J. An, K. Yang, J. Wu, N. Ye, S. Guo, Z. Liao, Achieving sustainable ultra-dense heterogeneous networks for 5g. IEEE Commun. Mag. **55**, 84–90 (2017)

11. K. Yang, N. Yang, N. Ye, M. Jia, Z. Gao, R. Fan, Non-orthogonal multiple access: achieving sustainable future radio access. IEEE Commun. Mag. **57**, 116–121 (2019)
12. N. Ye, X. Li, H. Yu, A. Wang, W. Liu, X. Hou, Deep learning aided grant-free noma toward reliable low-latency access in tactile internet of things. IEEE Trans. Indus. Inform. **15**, 2995–3005 (2019)
13. N. Ye, A. Wang, X. Li, W. Liu, X. Hou, H. Yu, On constellation rotation of noma with sic receiver. IEEE Commun. Lett. **22**, 514–517 (2018)
14. L. Zhang, W. Li, Y. Wu, X. Wang, S. Park, H.M. Kim, J. Lee, P. Angueira, J. Montalban, Layered-division-multiplexing: theory and practice. IEEE Trans. Broadcasting **62**, 216–232 (2016)
15. S.I. Park, Y. Wu, L. Zhang, J. Montalban, J. Lee, P. Angueira, S. Kwon, H.M. Kim, N. Hur, J. Kim, Low complexity layered division multiplexing system for the next generation terrestrial broadcasting, in *2015 IEEE International Symposium on Broadband Multimedia Systems and Broadcasting*, pp. 1–3 (2015)
16. J. Montalban, J. Barrueco, P. Angueira, L. Zhang, Y. Wu, W. Li, H. Kim, S. Park, J. Lee, Asynchronous n-layered division multiplexing (n-ldm), in *2016 IEEE International Symposium on Broadband Multimedia Systems and Broadcasting (BMSB)*, pp. 1–6 (2016)
17. J. Montalban, I. Angulo, C. Regueiro, Y. Wu, L. Zhang, S. Park, J. Lee, H.M. Kim, M. Velez, P. Angueira, Performance study of layered division multiplexing based on sdr platform. IEEE Trans. Broadcasting **61**, 436–444 (2015)
18. L. Zhang, Y. Wu, G.K. Walker, W. Li, K. Salehian, A. Florea, Improving lte *e* mbms with extended ofdm parameters and layered-division-multiplexing. IEEE Trans. Broadcasting **63**, 32–47 (2017)
19. E. Garro, J.J. Gimenez, S.I. Park, D. Gomez-Barquero, Layered division multiplexing with multi-radio-frequency channel technologies. IEEE Trans. Broadcasting **62**, 365–374 (2016)
20. Z. Zhang, Z. Ma, M. Xiao, Z. Ding, P. Fan, Full-duplex device-to-device-aided cooperative nonorthogonal multiple access. IEEE Trans. Veh. Technol. **66**, 4467–4471 (2017)
21. H.V. Vu, N.H. Tran, T. Le-Ngoc, Full-duplex device-to-device cellular networks: Power control and performance analysis. IEEE Trans. Veh. Technol. **68**, 3952–3966 (2019)
22. I. Krikidis, H.A. Suraweera, S. Yang, K. Berberidis, Full-duplex relaying over block fading channel: A diversity perspective. IEEE Trans. Veh. Technol. **11**, 4524–4535 (2012)
23. M. Byun, D. Park, B.G. Lee, Performance and distance spectrum of space-time codes in fast rayleigh fading channels. IEEE Trans. Commun. **56**, 2105–2115 (2008)
24. L. Jièmnez Rodrìguez, N.H. Tran, T. Le-Ngoc, Performance of full-duplex af relaying in the presence of residual self-interference. IEEE J. Sel. Areas Commun. **32**, 1752–1764 (2014)
25. G. Liu, W. Feng, Z. Han, W. Jiang, Performance analysis and optimization of cooperative full-duplex d2d communication underlaying cellular networks. IEEE Trans. Wirel. Commun. **18**, 5113–5127 (2019)
26. H. Chour, Y. Nasser, O. Bazzi, F. Bader, Full-duplex or half-duplex D2D mode? Closed form expression of the optimal power allocation, in *2018 25th International Conference on Telecommunications (ICT)*, pp. 498–504 (2018)
27. W. Chang, J. Teng, Energy efficient relay matching with bottleneck effect elimination power adjusting for full-duplex relay assisted d2d networks using mmwave technology. IEEE Access **6**, 3300–3309 (2018)
28. Y. Wang, Y. Niu, H. Wu, Z. Han, B. Ai, Q. Wang, Sub-channel allocation for device-to-device underlaying full-duplex mmwave small cells using coalition formation games. IEEE Trans. Veh. Technol. **68**, 11915–11927 (2019)
29. B. Ma, H. Shah-Mansouri, V. Wong, Full-duplex relaying for d2d communication in millimeter wave-based 5g networks. IEEE Trans. Wirel. Commun. **17**, 4417–4431 (2018)
30. R.T. Schwarz, T. Delamotte, K. Storek, A. Knopp, Mimo applications for multibeam satellites. IEEE Trans. Broadcasting **65**, 664–681 (2019)
31. W. Hong, K. Baek, Y. Lee, Y. Kim, S. Ko, Study and prototyping of practically large-scale mmwave antenna systems for 5g cellular devices. IEEE Commun. Mag. **52**, 63–69 (2014)

32. B. Yang, Z. Yu, J. Lan, R. Zhang, J. Zhou, W. Hong, Digital beamforming-based massive mimo transceiver for 5g millimeter-wave communications. IEEE Trans. Microwave Theory Techniques **66**, 3403–3418 (2018)
33. J.G. Andrews, T. Bai, M.N. Kulkarni, A. Alkhateeb, A.K. Gupta, R.W. Heath, Modeling and analyzing millimeter wave cellular systems. IEEE Trans. Commun. **65**, 403–430 (2017)
34. X. Gao, D. Niyato, P. Wang, K. Yang, J. An, Contract design for time resource assignment and pricing in backscatter-assisted RF-powered networks. IEEE Wirel. Commun. Lett., 1 (2019)
35. X. Gao, P. Wang, D. Niyato, K. Yang, J. An, Auction-based time scheduling for backscatter-aided RF-powered cognitive radio networks. IEEE Trans. Wirel. Commun. **18**, 1684–1697 (2019)
36. Z. Wang, Q. Liu, M. Li, W. Kellerer, Energy efficient analog beamformer design for mmwave multicast transmission. IEEE Trans. Green Commun. Netw. **3**, 552–564 (2019)
37. B. Wang, L. Dai, Z. Wang, N. Ge, S. Zhou, Spectrum and energy-efficient beamspace mimo-noma for millimeter-wave communications using lens antenna array. IEEE J. Sel. Areas Commun. **35**, 2370–2382 (2017)
38. J. Jin, C. Xiao, W. Chen, Y. Wu, Hybrid precoding in mmwave mimo broadcast channels with dynamic subarrays and finite-alphabet inputs, in *2018 IEEE International Conference on Communications (ICC)*, pp. 1–6 (2018)
39. T. Bai, R.W. Heath, Coverage and rate analysis for millimeter-wave cellular networks. IEEE Trans. Wirel. Commun. **14**, 1100–1114 (2015)
40. S.A.R. Naqvi, S.A. Hassan, Combining noma and mmwave technology for cellular communication,' in *2016 IEEE 84th Vehicular Technology Conference (VTC-Fall)*, pp. 1–5 (2016)
41. Y. Sun, Z. Ding, X. Dai, On the performance of downlink noma in multi-cell mmwave networks. IEEE Commun. Lett. **22**, 2366–2369 (2018)
42. N. Deng, M. Haenggi, A fine-grained analysis of millimeter-wave device-to-device networks. IEEE Trans. Commun. **65**, 4940–4954 (2017)
43. Y. Huang, F. Al-Qahtani, C. Zhong, Q. Wu, J. Wang, H. Alnuweiri, Performance analysis of multiuser multiple antenna relaying networks with co-channel interference and feedback delay. IEEE Trans. Commun. **62**, 59–73 (2014)

Chapter 3
Efficient Detection: Asynchronous Multi-user Detection for Code Domain NOMA

In this chapter, we focus on asynchronous code-domain NOMA detection. Section 3.1 presents the background and motivation of the research on asynchronous code-domain NOMA detection. Section 3.2 presents the system model and the receiver model of asynchronous code-domain NOMA. Section 3.3 constructs the 3D factorgraph and deploys two message passing mechanisms. Section 3.4 introduces the expectation propagation mechanism, designs the 3D-EPA in the proposed 3D factorgraph, and develops the state evolution to analyze the performance of the 3D-EPA. Section 3.5 conducts the link-level simulation results and complexity analysis. Section 3.6 concludes the work of the chapter.

3.1 Introduction

By the next decade, the worldwide Internet of Things equipment will hit an average of 10 million access per square kilometer to support services in a wide area [1, 2]. The ubiquitous access requirements of massive devices will be a strong challenge for wireless communications [3]. To this end, NOMA is proposed as an essential enabling technology to accommodate massive connections for 5G and beyond by non-orthogonal superposition over the same time-frequency resources [4–9]. As a major branch of NOMA technologies, code-domain NOMA exploits the judiciously designed spreading codes as well as the advanced multi-user detector to suppress the IUI triggered by superposition transmissions. Compared with the orthogonal transmission schemes, NOMA can achieve dramatically higher overloading and spectral efficiency with the same physical resources [10, 11].

However, for application scenarios with large channel dynamics, e.g., nonterrestrial wireless communications [12, 13], or limited scheduling capabilities, e.g., grant-free communications [14], the strict synchronization among non-orthogonal users becomes cost-worthy and impractical. In these cases, multiple non-orthogonal

J. Li et al., *Key Technologies of High Frequency Wireless Communications*,
https://doi.org/10.1007/978-981-96-5894-7_3

signals can be asynchronously superposed over the transmission channel [15]. The resulting asynchronous IUI is far more complex than that of traditional NOMA, which becomes an unresolved challenge to the multi-user detection for asynchronous NOMA.

3.1.1 Literature and Motivation

NOMA relies on the ingenious designs of the transmitter and receiver methods to achieve efficient multi-user transmission. At the transmitter, power differences [16] and user-specific spreading codes with strictly sparse [17] or low correlation [18] are designed to distinguish users. At the receiver, several multi-user detection schemes are designed by utilizing multiple access signatures and interference patterns to alleviate the IUI [19].

The existing multi-user detection algorithms mainly focus on synchronous transmissions. These methods can be classified into two categories, i.e. the non-iteration-based approaches and the iteration-based approaches. The non-iteration-based approaches include the minimum mean square error-successive interference cancellation (MMSE-SIC) [20], the zero-forcing (ZF) [21], etc. As a typical method, the MMSE-SIC [20] treats the interference as noise and performs estimation user-by-user directly. While this feature facilitates its extension to the asynchronous transmission systems, its exploitation of the interference structure is still inadequate. The iteration-based algorithms mainly include MPA [22], EM-MPA [23], AMP [24, 25], EPA [26–32], etc. They are all designed on the loopy factor-graph which perform symbol estimation using message passing iteratively. Among them, the MPA has excellent performance close to maximum likelihood detection. To address the exponential complexity introduced by MPA, EM-MPA, AMP and EPA deploy approximation mechanisms in the belief propagation algorithm. In particular, the EPA maps the complex distribution to a Gaussian distribution using the moment matching method, which sharply reduces the complexity from exponential level to linear level. However, these iteration-based methods are all designed for synchronous NOMA, and therefore cannot characterize and mitigate the asynchronous IUI.

To accomplish asynchronous multi-user detection, some efforts have been made for different NOMA schemes. The conventional SIC technique was extended to asynchronous power-domain NOMA system with the detection orders of user-by-user [33] and symbol-by-symbol [34]. A trellis-based detection algorithm was also provided in [33]. Besides, a T-SIC technology was proposed in [35], which develops a triangular pattern to perform interference cancellation, and further optimizes detection performance through an iterative mechanism. Meanwhile, the factor-graph representation was studied in [36, 37], where elementary signal estimation (ESE) is deployed for low-complexity asynchronous multiuser detection. However, the above mentioned algorithms do not exploit the correlation among different spreading codes, and thus are not suitable for code-domain NOMA. In the asynchronous sparse code multiple access (SCMA) systems, the authors in [38] proposed three algorithms, i.e., the belief propagation message passing algorithm (BP-MPA), the belief propagation with equal gain combining message passing algorithm (BP-EGC-MPA), and partial phase-coherent algorithm. Nevertheless, these methods are difficult to be applied in the code-domain NOMA with dense spreading codes, e.g., the Welch-bound equality (WBE)-based codes [18], due to the high-complexity of the MPA. In conclusion, existing asynchronous algorithms cannot perform multi-user detection efficiently for asynchronous code-domain NOMA with dense spreading codes.

3.1.2 Contributions

In this chapter, we propose a novel low-complexity multi-user detection algorithm for symbol-asynchronous code-domain NOMA. Considering that the IUI is extended to the surrounding time intervals, we introduce time delay as a new dimension into the traditional factor-graph model. Since it is non-trivial to perform symbol estimation directly based on this model, we consider a novel representation of the 3D factor-graph by dividing it into several slices. On this basis, we propose a low-complexity expectation propagation mechanism which iterates between intra-slice estimation, by projecting the asynchronous interference via Gaussian approximation, and inter-slice message passing, by propagating the previous estimates among the users to refine the approximation. The main contributions of this chapter are summarized as follows:

- Firstly, we construct a novel factor-graph model for code-domain NOMA to simultaneously characterize symbol spreading and asynchronous interference. In this model, time delay is introduced as an additional dimension in the traditional bipartite factor-graph. Then, the resulting 3D factor-graph is divided into several coupled sub-factor-graphs with each aligned with one NOMA symbol of a specific user. Meanwhile, two message passing mechanisms, i.e. intra-slice message passing and inter-slice message passing, are introduced to respectively characterize the probabilistic relationships within and between sub-factor-graphs.
- Secondly, we develop an expectation propagation mechanism on the 3D factor-graph and propose a novel low-complexity multi-user detection algorithm, namely the 3D-EPA. The proposed algorithm iterates between per-user estimation and inter-user message passing. It projects the complex symbol estimation onto a Gaussian distribution and performs symbol estimation with the assistance of the information passed from adjacent sub-factor-graphs. In addition, we also extend the 3D-EPA to multiple antenna systems.
- Finally, we develop state evolution analysis and perform Monte Carlo simulations to verify the BER performance of the 3D-EPA with asynchronous NOMA. The simulation results show that the proposed asynchronous code-domain NOMA with the 3D-EPA outperforms the synchronous counterpart especially under high overloadings, and is also superior to straightforward typical asynchronous detection algorithms. Complexity and convergence are also analyzed.

3.2 System Model

3.2.1 *Uplink Asynchronous Code-Domain NOMA System Model*

Consider an asynchronous uplink system for code-domain NOMA with K independent users and one base station (BS), as shown in Fig. 3.1. The users and the BS are equipped with a single antenna. In each frame, the k-th user aims to transmit the binary information bit vector $\mathbf{b}_k = [b_k^1, ..., b_k^i, ..., b_k^L]^T$, where L is the frame length. First, $\mathbf{b}_k$ is modulated to $\mathbf{c}_k = [c_k^1, ..., c_k^i, ..., c_k^L]^T$ via a binary phase-shift keying modulator, where $[\cdot]^T$ represents the transpose of a vector or matrix. Then each modulation symbol c_k^i, $1 \leq i \leq L$, is spread by a user-specific spreading sequence $\mathbf{w}_k$ to generate a codeword $\mathbf{x}_k^i$, which is given by

$$\mathbf{x}_k^i = c_k^i \cdot \mathbf{w}_k = [x_{k,1}^i, ..., x_{k,n}^i, ..., x_{k,N}^i]^T, \tag{3.1}$$

where N is the spreading factor representing the number of occupied orthogonal physical resources. The system overloading can be given by $\eta = K/N \times 100\%$.

To generate the continuous transmit waveform signal, the transmitted symbol matrix of the k-th user, defined as $\mathbf{X}_k = [\mathbf{x}_k^1, ..., \mathbf{x}_k^i, ..., \mathbf{x}_k^L]^T$, is then convolved with the pulse shaping waveform $s(t)$ which has unit energy. The transmitted signal of the k-th user on the n-th orthogonal resource is given by

$$x_{k,n}(t) = \sum_{i=1}^{L} x_{k,n}^i s\,(t - (i-1)T), \tag{3.2}$$

where T denotes the time interval of each symbol.

In the asynchronous transmission system, each user's signal reaches the receiver with a user-specific time delay. We assume that the time delay is limited to one symbol interval T. This is because a larger time delay can be eliminated by frame synchronization, which is easier to achieve than symbol synchronization [39]. We denote τ_k as the time delay of the k-th user, and set $0 < \tau_1 < ... < \tau_K < T$ without loss of generality. Then the received signal on the n-th orthogonal resource can be expressed as

$$y_n(t) = \sum_{k=1}^{K} h_{k,n} x_{k,n}(t - \tau_k) + z_n(t), \tag{3.3}$$

where $h_{k,n}$ represents the channel coefficient of the k-th user on the n-th resource element, and $z_n(t)$ represents the noise on $y_n(t)$. At the receiver side, a bank of matched filters are deployed to generate K discrete detected symbol sequences for the K users respectively. For the k-th user, $y_n(t)$ is first aligned to its specific time delay, and then matched-filtered and downsampled to generate the discrete symbol $y_{k,n}^i$. This process can be given by

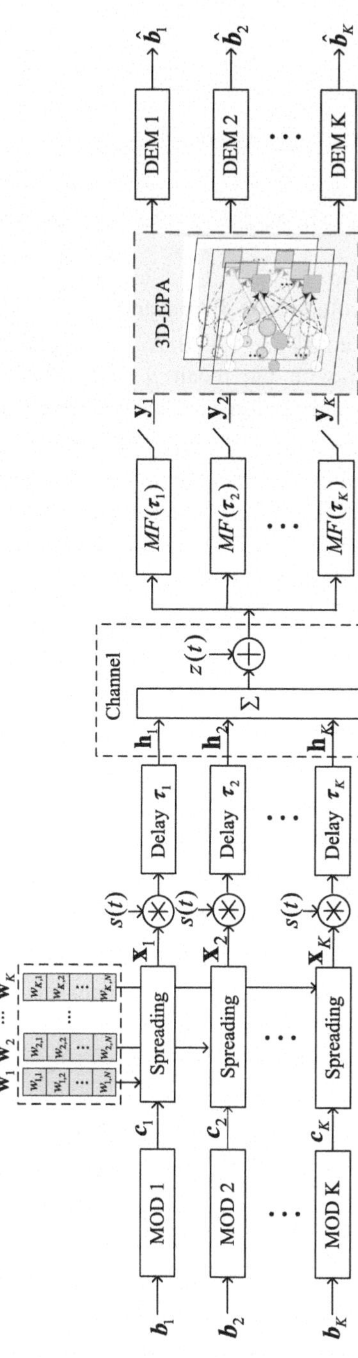

Fig. 3.1 Symbol-asynchronous code-domain NOMA system

$$
\begin{aligned}
y_{k,n}^i &= \left[y_n(t) * MF(\tau_k) \right]_{t=(i-1)T+\tau_k}^{iT+\tau_k} \\
&= \int_{(i-1)T+\tau_k}^{iT+\tau_k} y_n(t) s\left(t - (i-1)T - \tau_k\right) dt \\
&= h_{k,n} x_{k,n}^i + \sum_{m \neq k} h_{m,n} \rho_{m,k} x_{m,n}^i \\
&\quad + \sum_{m<k} h_{m,n} v_{m,k} x_{m,n}^{i+1} + \sum_{m>k} h_{m,n} v_{m,k} x_{m,n}^{i-1} + z_{k,n}^i,
\end{aligned}
\tag{3.4}
$$

where $*$ denotes the convolution operation and $z_{k,n}^i$ is additive white gaussian noise (AWGN) on $y_{k,n}^i$, i.e., $z_{k,n}^i \sim \mathcal{CN}(0, N_0)$, and $\mathcal{CN}(\mu, \xi)$ denotes a complex Gaussian distribution with the mean μ and variance ξ. The terms $\rho_{m,k}$ and $v_{m,k}$ denote the cross-correlation coefficients which are defined as $\rho_{m,k} = \int_0^T s(t - |\tau^m - \tau^k|)s(t)dt$ and $v_{m,k} = \int_0^T s(t - (T - |\tau^m - \tau^k|))s(t)dt$, where $\rho_{m,k}$ represents the cross-correlation coefficient between symbols of the m-th and the k-th user in a same symbol interval, and $v_{m,k}$ represents the cross-correlation coefficient between symbols of the m-th and the k-th user in adjacent symbol intervals.

It can be observed in (3.4) that a complex interference pattern is introduced by asynchronous non-orthogonal transmission, which is also shown in Fig. 3.2a. For example, the IUI of $x_{k,n}^i$ not only comes from other users' transmitted symbols in the same symbol interval, i.e., $x_{m,n}^i$, $m \neq k$, but also from the surrounding symbol intervals of other users, i.e., $x_{m,n}^{i+1}$, for $m < k$, and $x_{m,n}^{i-1}$, for $m > k$. This greatly increases the difficulty of interference suppression compared with its synchronous counterpart.

3.2.2 Receiver Model

Consider an iterative multi-user detector with the one-shot architecture. Assume that the receiver knows the ideal time delay[1] τ_k and channel state information $h_{k,n}$, $\forall k, n$. Accordingly, the multi-user detector generates soft estimates of the users' transmitted symbols, i.e., the extrinsic log-likelihood ratio (LLR) $\Lambda\left(b_k^i\right)$, which is given by

$$
\Lambda\left(b_k^i\right) = \log \frac{\sum_{\mathbf{b}_k \in \mathcal{B}_{k,i}^+} p\left(\mathbf{b}_k \mid \mathbf{Y}\right)}{\sum_{\mathbf{b}_k \in \mathcal{B}_{k,i}^-} p\left(\mathbf{b}_k \mid \mathbf{Y}\right)},
\tag{3.5}
$$

---

[1] Time delay can be obtained by channel estimation. For example, we can insert a pilot before each user's data frame, and obtain the time delay by estimating the correlation peak of that pilot.

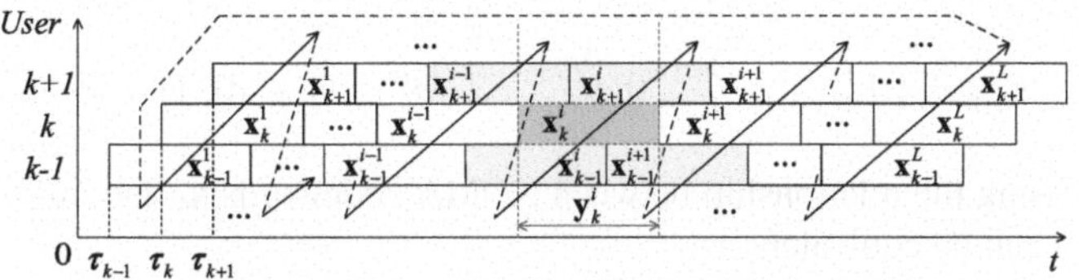

(a) Received signal structure and detection order.

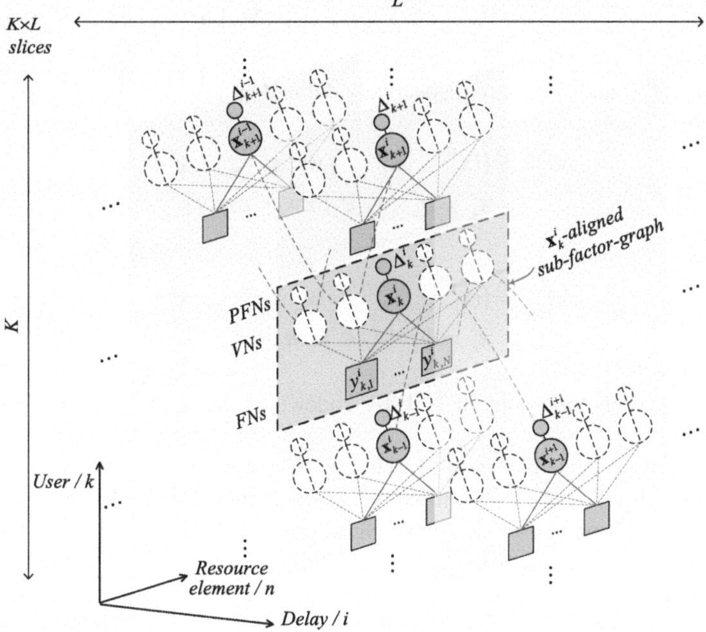

(b) 3D factor-graph representation for symbol-asynchronous
code-domain NOMA.

Fig. 3.2 Signal model and 3D factor-graph model

where $\mathbf{Y}$ is the collection of $\mathbf{y}_k^i = [y_{k,1}^i, ..., y_{k,n}^i, ..., y_{k,N}^i]^T$, $\forall i, k$, defined as

$$\mathbf{Y} = \begin{bmatrix} \mathbf{y}_1^1 & \cdots & \mathbf{y}_1^L \\ \vdots & \ddots & \vdots \\ \mathbf{y}_K^1 & \cdots & \mathbf{y}_K^L \end{bmatrix},$$
(3.6)

and the terms $\mathcal{B}_{k,i}^+$ and $\mathcal{B}_{k,i}^-$ are respectively defined as

$$\mathcal{B}_{k,i}^+ \triangleq \{(b_k^1, \ldots, b_k^{i-1}, 1, b_k^{i+1}, \ldots, b_k^L) : b_k^r \in \{0, 1\}, r \neq i\},$$
(3.7)

and

$$\mathcal{B}_{k,i}^{-} \triangleq \{(b_k^1, \ldots, b_k^{i-1}, 0, b_k^{i+1}, \ldots, b_k^L) : b_k^r \in \{0, 1\}, r \neq i\}. \tag{3.8}$$

Then, by using the relationship between NOMA codewords $\mathbf{x}_k$ and information bits $\mathbf{b}_k$, $\Lambda\left(b_k^i\right)$ can be equivalent as

$$\Lambda\left(\mathbf{x}_k^i\right) = \log \frac{\sum_{\mathbf{X}_k \in \mathcal{M}_k^{i,+}} p\left(\mathbf{X}_k \mid \mathbf{Y}\right)}{\sum_{\mathbf{X}_k \in \mathcal{M}_k^{i,-}} p\left(\mathbf{X}_k \mid \mathbf{Y}\right)}. \tag{3.9}$$

The terms $\mathcal{M}_k^{i,+}$ and $\mathcal{M}_k^{i,-}$ are respectively defined as

$$\begin{aligned} \mathcal{M}_k^{i,+} \triangleq \{&(\mathbf{x}_k^1, \ldots, \mathbf{x}_k^{i-1}, +\mathbf{w}_k, \mathbf{x}_k^{i+1}, \ldots, \mathbf{x}_k^L) : \\ &\mathbf{x}_k^r = b_k^r \cdot \mathbf{w}_k \in \{\pm \mathbf{w}_k\}, r \neq i\}, \end{aligned} \tag{3.10}$$

and

$$\begin{aligned} \mathcal{M}_k^{i,-} \triangleq \{&(\mathbf{x}_k^1, \ldots, \mathbf{x}_k^{i-1}, -\mathbf{w}_k, \mathbf{x}_k^{i+1}, \ldots, \mathbf{x}_k^L) : \\ &\mathbf{x}_k^r = b_k^r \cdot \mathbf{w}_k \in \{\pm \mathbf{w}_k\}, r \neq i\}, \end{aligned} \tag{3.11}$$

which are sets of code-domain NOMA codewords mapped from $\mathcal{B}_{k,i}^{+}$ and $\mathcal{B}_{k,i}^{-}$, respectively.

However, direct calculation of (3.9) using the maximum a posteriori probability detection result in $\mathcal{O}(M^{2K-1})$ complexity which is intolerable in practice. Furthermore, traditional multi-user detection algorithms, such as message passing algorithm (MPA), expectation propagation algorithm (EPA), and so on are all based on factor-graphs designed for synchronous transmissions. They cannot characterize the asynchronous interference of code-domain NOMA due to the additional IUI which comes from the surrounding time intervals. Therefore, a low-complexity multi-user detection algorithm for asynchronous transmission should be studied.

3.3 3D Factor-Graph and Message Passing Mechanisms

This section aims to characterize the asynchronous IUI using the factor-graph model. Specifically, we develop a novel factor-graph model by introducing time delay as a new dimension into the traditional factor-graph. Then, to characterize the interference pattern in an intuitive way, a 3D factor-graph is then constructed by dividing the above-mentioned model into several slices. Moreover, two message passing mechanisms are developed to describe the relationships between those nodes in 3D factor-graph.

3.3.1 3D Factor-Graph Representation

We start with the joint distribution of the transmitted and received signals of the multi-user systems. To illustrate the interference pattern of NOMA, the joint distribution can be factorized and modeled as a factor-graph according to the principle of the probabilistic graphical model [40]. The traditional factor-graph for synchronous NOMA is developed to characterize the signal model of symbols within a single time interval [22, 26]. However, it fails to characterize the asynchronous transmission systems, since the IUI, in this case, also comes from the surrounding time intervals.

To tackle this problem, we consider the joint distribution of symbols in the entire frame. We define $\mathbf{X} = [\mathbf{X}_1, \ldots, \mathbf{X}_k, \ldots, \mathbf{X}_K]^T$ as the collection of the transmitted symbol matrices of all K users in the entire frame. Then, the joint distribution $P(\mathbf{X}, \mathbf{Y})$ can be factorized as

$$
\begin{aligned}
P(\mathbf{X}, \mathbf{Y}) &= \prod_{i=1}^{L}\prod_{k=1}^{K} P(\mathbf{X}, \mathbf{y}_k^i) = \prod_{i=1}^{L}\prod_{k=1}^{K} P(\mathbf{y}_k^i|\mathbf{X})P(\mathbf{X}) \\
&\overset{(a)}{=\!=} \prod_{i=1}^{L}\prod_{k=1}^{K}\underbrace{\left(\prod_{n=1}^{N} P(y_{k,n}^i|\mathbf{X})\prod_{s=1}^{L}\prod_{r=1}^{K} P(\mathbf{x}_r^s)\right)}_{\text{Three-dimensional: Resource element, user, and time delay}} .
\end{aligned}
\tag{3.12}
$$

Considering that $y_{k,n}^i$, and $x_{k,n}^i$, $\forall k, i, n$, are independently and identically distributed, $P(\mathbf{X}, \mathbf{Y})$ can be firstly factorized as the product of KL joint distributions $P(\mathbf{X}, \mathbf{y}_k^i)$, $\forall k, i$. Further, by factorizing $\mathbf{X}$ and $\mathbf{y}_k^i$, step (a) can be obtained. We observe that besides the two dimensions representing resource element (indexed by n) and user (indexed by k or r), time delay (indexed by i or s) is also introduced as a new dimension.

According to step (a) of (3.12), a factor-graph model can be constructed with three dimensions, i.e., resource element dimension (n), user dimension $(k$ or $r)$, and time delay dimension $(i$ or $s)$, to characterize the factorization of $P(\mathbf{X}, \mathbf{Y})$. However, this factor-graph representation is not fully intuitive to identify the interference pattern of each transmitted symbol. Moreover, high computational complexity has to be tolerated if we perform multi-user detection directly on this model. Therefore, we divide the factor-graph according to each symbol of all users, which results in KL slices. Then, the proposed 3D factor-graph $\mathcal{G}(\mathcal{V}, \mathcal{F}, \mathcal{PF}, \mathcal{E})$ is obtained.

As shown in Fig. 3.2b, $\mathcal{G}(\mathcal{V}, \mathcal{F}, \mathcal{PF}, \mathcal{E})$ consists of VN collection $\mathcal{V}$, function node (FN) collection $\mathcal{F}$, prior factor node (PFN) collection $\mathcal{PF}$, and the edge collection $\mathcal{E}$, each of which represents the transmitted symbols, the likelihood functions, the prior probabilities, and the messages passing between two nodes, respectively. Meanwhile, each slice in this 3D factor-graph is aligned with only one transmitted symbol to represent the IUI suffered by this symbol. Here, we name the slice which is aligned with $\mathbf{x}_k^i$ as the $\mathbf{x}_k^i$-aligned sub-factor-graph.

The $\mathbf{x}_k^i$-aligned sub-factor-graph, as shown in Fig. 3.2b, contains three types of nodes, i.e., K VNs, N FNs, and K PFNs. Among these VNs, the k-th VN represents the transmitted symbol $\mathbf{x}_k^i$, the q-th VN, $q \neq k$, represents the interference from the q-th user on the transmitted symbol $\mathbf{x}_k^i$. It contains two parts to represent two interfering symbols, since two adjacent symbols of the q-th user interfere with $\mathbf{x}_k^i$ in asynchronous transmission. We denote $V_q^{[\mathbf{x}_k^i]}$ as an index set to represent the q-th user's symbols which interfere with $\mathbf{x}_k^i$, and use $\mathbf{x}_q^u$, $u \in V_q^{[\mathbf{x}_k^i]}$, to represent these interfering symbols. Here, $V_q^{[\mathbf{x}_k^i]}$ is equals to $\{i, i+1\}$ for $q < k$, or $\{i-1, i\}$ for $q > k$, and $|V_q^{[\mathbf{x}_k^i]}|$ is defined as the number of indexes. For example, the q-th VN, $q < k$, in the $\mathbf{x}_k^i$-aligned sub-factor-graph represents the interference from the q-th user to the k-th user, which is composed of $|V_q^{[\mathbf{x}_k^i]}| = 2$ parts for $i < L$, representing $\mathbf{x}_q^i$ and $\mathbf{x}_q^{i+1}$, respectively. As for the FNs and PFNs in the $\mathbf{x}_k^i$-aligned sub-factor-graph, the n-th FN represents the likelihood function $P(y_{k,n}^i | \mathbf{X})$, and PFN Δ_k^i represents the prior probability of $\mathbf{x}_k^i$.

Besides, as shown in Fig. 3.2b, we identify two types of edges in the 3D factor-graph to respectively describe the relationship among VNs and FNs. In the $\mathbf{x}_k^i$-aligned sub-factor-graph, one type of edges which are marked as the red-solid lines connect FNs with the k-th VN. They are used to describe the probabilistic relationship between the two connected nodes. The other type of edges which are marked as the blue-dashed lines are those related to the q-th VN. They can be further divided into two categories. For the one that connects two VNs in two adjacent sub-factor-graphs, it is used to represent the relationship between two identical transmitted symbols in different sub-factor-graphs. For the one that connects the q-th VN with a FN, it is used to describe the probabilistic relationship between the two connected nodes.

3.3.2 Message Passing Mechanisms over 3D Factor-Graph

In this subsection, we develop two message passing mechanisms for the above-mentioned two types of edges, respectively. Without loss of generality, we focus on the slice aligned with $\mathbf{x}_k^i$, i.e., the $\mathbf{x}_k^i$-aligned sub-factor-graph, as shown in Fig. 3.3. For the red-solid edges which connect FNs with the k-th VN, we develop the intra-slice message passing mechanism. For the other blue-dashed edges, we develop the inter-slice message passing mechanism. In the following, we explain these two message passing mechanisms, respectively.

3.3.2.1 Intra-Slice Message Passing

Intra-slice message passing deploys the sum-product algorithm on the edges which connect FNs with the k-th VN to update the information of these nodes. As shown at the top of Fig. 3.3a, we define $I_{k \to f_n}^{i,[\mathbf{x}_k^i]}$ as the message passed from the k-th VN to the

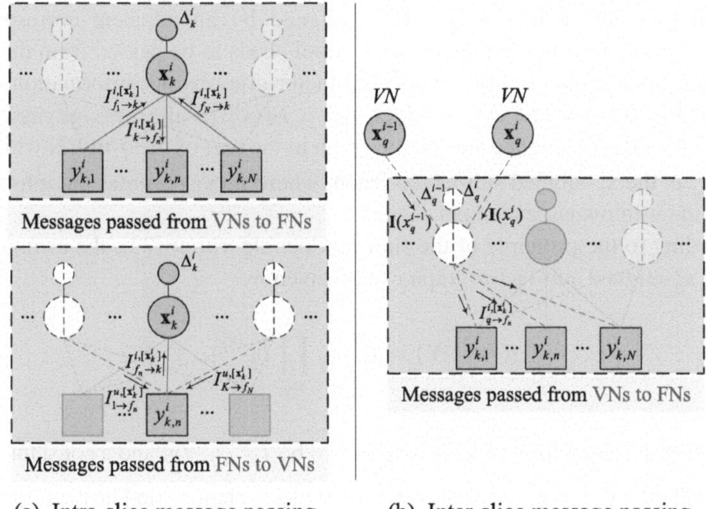

Fig. 3.3 Two message passing mechanisms over 3D factor-graph

n-th FN in the $\mathbf{x}_k^i$-aligned sub-factor-graph. Similarly, define $I_{f_n \to k}^{i,[\mathbf{x}_k^i]}$ as the message passed from the n-th FN to the k-th VN in the $\mathbf{x}_k^i$-aligned sub-factor-graph. According to the principle of sum-product algorithm, $I_{k \to f_n}^{i,[\mathbf{x}_k^i]}$ can be calculated based on $I_{f_n \to k}^{i,[\mathbf{x}_k^i]}$ which are passed from all but the n-th FNs to the k-th VN. This process can be expressed as

$$I_{k \to f_n}^{i,[\mathbf{x}_k^i]} = I_{\Delta_k^i \to k} \prod_{m \neq n} I_{f_m \to k}^{i,[\mathbf{x}_k^i]}, \tag{3.13}$$

where $I_{\Delta_k^i \to k}$ denotes the prior probability of $\mathbf{x}_k^i$, which can be assigned to 1 if no external information is fed back from the decoder to the multi-user detector.

Then, as shown at the bottom of Fig. 3.3a, for the message $I_{f_n \to k}^{i,[\mathbf{x}_k^i]}$, it is composed of $I_{q \to f_n}^{u,[\mathbf{x}_k^i]}$, $\forall q \neq k$, $\forall u \in V_q^{[\mathbf{x}_k^i]}$ and can be given by

$$I_{f_n \to k}^{i,[\mathbf{x}_k^i]} = \sum_{q \in n(k)}^{\mathbf{x}^q \neq \mathbf{x}^k} P(y_{k,n}^i | \mathbf{X}) \prod_{q \in n(k) \setminus k} \prod_{u} I_{q \to f_n}^{u,[\mathbf{x}_k^i]}, \tag{3.14}$$

where $n(k)$ denotes a set of users that interfere with the k-th user.

3.3.2.2 Inter-Slice Message Passing

Inter-slice message passing aims to update the information on the q-th VN, $q \neq k$, as well as its message to FNs, as shown in Fig. 3.3b. Since each of the symbols represented by the q-th VN, $q \neq k$, is aligned by an adjacent corresponding sub-factor-graph, information about this symbol needs to be passed from the corresponding sub-factor-graph to the current sub-factor-graph as side information. Therefore, we define $\mathbf{I}(\mathbf{x}_q^u) = [I_1(\mathbf{x}_q^u), \ldots, I_n(\mathbf{x}_q^u), \ldots, I_N(\mathbf{x}_q^u)]$ as the message passed from the q-th VN in the $\mathbf{x}_q^u$-aligned sub-factor-graph to the part of the q-th VN which represents $\mathbf{x}_q^u$ in the $\mathbf{x}_k^i$-aligned sub-factor-graph, where $I_n(\mathbf{x}_q^u)$ denotes the information of $\mathbf{x}_q^u$ on the n-th resource element.

According to the principle of the sum-product algorithm [22], the estimation of $\mathbf{x}_q^u$ in the $\mathbf{x}_q^u$-aligned sub-factor-graph can be given by

$$p\left(\mathbf{x}_q^u \mid \mathbf{Y}\right) = I_{\Delta_q^u \to q} \prod_{\forall m} I_{f_m \to q}^{u,[\mathbf{x}_q^u]}. \tag{3.15}$$

Observe that the estimation of $\mathbf{x}_q^u$ is determined by $I_{f_m \to q}^{u,[\mathbf{x}_q^u]}$, $\forall m$ and a constant $I_{\Delta_q^u \to q}$, we can define $I_n(\mathbf{x}_q^u) = I_{f_n \to q}^{u,[\mathbf{x}_q^u]}$ as the estimation of $\mathbf{x}_q^u$. Then, to update the information on FNs in the $\mathbf{x}_k^i$-aligned sub-factor-graph, we define $I_{q \to f_n}^{u,[\mathbf{x}_k^i]}$ as the message passed from the part of the q-th VN which represents $\mathbf{x}_q^u$ to the n-th FN in the $\mathbf{x}_k^i$-aligned sub-factor-graph. Thus, we can utilize $I_n(\mathbf{x}_q^u)$ to represent $I_{q \to f_n}^{u,[\mathbf{x}_k^i]}$, i.e.,

$$I_{q \to f_n}^{u,[\mathbf{x}_k^i]} = I_n(\mathbf{x}_q^u). \tag{3.16}$$

Remark 3.1 A novel 3D factor-graph is constructed to characterize asynchronous IUI based on the entire frame. In this model, each slice only estimates the aligned symbol by using intra-slice message passing. Meanwhile, inter-slice message passing is also introduced to pass other symbols' estimates as side information to assist estimation in the current slice. However, deploying the sum-product algorithm directly in the 3D factor-graph will still result in unbearable complexity due to the excessive VNs which represent the interfering symbols. In addition, an iterative mechanism suitable for the proposed 3D factor-graph is required.

3.4　Proposed 3D-EPA for Code-Domain NOMA

In this section, the expectation propagation mechanism is introduced on the 3D factor-graph, which inspires a low-complexity multi-user detection algorithm, namely the 3D-EPA. In this algorithm, the sum-product algorithm is simplified using the Gaussian approximation. Meanwhile, a frame-level iterative method is developed for the proposed 3D-EPA.

3.4.1 Expectation Propagation Mechanism

Expectation propagation is a low-complexity and efficient inference method based on Bayesian approximation [41]. It uses a moment matching method mapping a general distribution to a distribution composed of exponential families. Mathematically, it can be explained by Kullback-Leibler (K-L) divergence, which is a method of describing the degree of fitting between approximate distribution $\tilde{p}$ and real distribution p. Then, the approximate distribution $\tilde{p}$ can be given by

$$\tilde{p} = \text{proj}_{\mathbb{Q}}(p) = \arg\min_{\tilde{p}\in\mathbb{Q}} KL(p||\tilde{p}), \tag{3.17}$$

where $\text{proj}_{\mathbb{Q}}(\cdot)$ denotes a projection function that maps a real distribution into some distribution set $\mathbb{Q}$, and $KL(p||\tilde{p})$ denotes the K-L divergence. Here, the minimum K-L divergence is obtained by matching the expectation of sufficient statistics, i.e., moment matching. Intuitively, the real distribution p is mapped to a family of Gaussian distributions. Given a loopy factor-graph, the posterior probability of each symbol can be calculated using the expectation propagation mechanism. Therefore, according to the message update rules of expectation propagation, (3.13) and (3.14) can be respectively given by

$$I_{k\to f_n}^{i,[\mathbf{x}_k^i]} = \frac{\text{proj}_{\mathbb{Q}}(p(\mathbf{x}_k^i))}{I_{f_n\to k}^{i,[\mathbf{x}_k^i]}}, \tag{3.18}$$

and

$$I_{f_n\to k}^{i,[\mathbf{x}_k^i]} = \frac{\text{proj}_{\mathbb{Q}}(q_{f_n}(\mathbf{x}_k^i))}{I_{k\to f_n}^{i,[\mathbf{x}_k^i]}}. \tag{3.19}$$

Here, $p(\mathbf{x}_k^i)$ in (3.18) represents the estimated probability distribution function of $\mathbf{x}_k^i$, which is given by

$$p(\mathbf{x}_k^i) = I_{\Delta_k^i\to k} \prod_{\forall m} I_{f_m\to k}^{i,[\mathbf{x}_k^i]}, \tag{3.20}$$

and $q_{f_n}(\mathbf{x}_k^i)$ in (3.19) represents the joint distribution function of $y_{k,n}^i$ as well as all the transmitted symbols related to $y_{k,n}^i$, which is given by

$$q_{f_n}(\mathbf{x}_k^i) = \sum_{q\in n(k)} P(y_{k,n}^i|\mathbf{X}) \prod_{q\in n(k)} \prod_{u} I_{q\to f_n}^{u,[\mathbf{x}_k^i]}. \tag{3.21}$$

Based on the properties of the Gaussian distribution, all terms in (3.18) and (3.19) can be written as a Gaussian distribution, which transforms the complex message updating into the calculation of the mean and variance.

3.4.2 Expectation Propagation over 3D Factor-Graph

In this subsection, an iterative multi-user detection algorithm, namely the 3D-EPA, is proposed based on the 3D factor-graph for asynchronous code-domain NOMA. The 3D-EPA operates on the entire 3D factor-graph along the time delay dimension. For each sub-factor-graph, the 3D-EPA deploys the expectation propagation mechanism to estimate the aligned symbol with the assistance of messages passed from adjacent sub-factor-graphs. The proposed 3D-EPA is detailed as follows.

We focus on the message passing in the $\mathbf{x}_k^i$-aligned sub-factor-graph in the following. Considering that the symbol estimates and the messages are approximated as Gaussian distributions according to the principle of the EPA [41], we denote $P^{i_{\text{iter}}}(x_{k,n}^i)$ as the estimated symbol belief of $x_{k,n}^i$ in the i_{iter}-th iteration, which can be given by

$$P^{i_{\text{iter}}}(x_{k,n}^i) \propto \mathcal{CN}(\mu_{k,n}^{i,i_{\text{iter}}}, \xi_{k,n}^{i,i_{\text{iter}}}), \tag{3.22}$$

where $\mu_{k,n}^{i,i_{\text{iter}}}$ and $\xi_{k,n}^{i,i_{\text{iter}}}$ respectively represent the mean and the variance of the approximated Gaussian distribution of $x_{k,n}^i$ in the i_{iter}-th iteration. For notation simplicity, we omit the superscript i_{iter} of all messages in the following.

Then, $I_{f_n \to k}^{i,[\mathbf{x}_k^i]}$ and $I_{q \to f_n}^{j,[\mathbf{x}_k^i]}, q \neq k, j \in V_q^{[\mathbf{x}_k^i]}$ in the $\mathbf{x}_k^i$-aligned sub-factor-graph can be respectively given by

$$I_{f_n \to k}^{i,[\mathbf{x}_k^i]} \propto \mathcal{CN}(\mu_{f_n \to k}^{i,[\mathbf{x}_k^i]}, \xi_{f_n \to k}^{i,[\mathbf{x}_k^i]}), \tag{3.23}$$

and

$$I_{q \to f_n}^{j,[\mathbf{x}_k^i]} \propto \mathcal{CN}(\mu_{q \to f_n}^{j,[\mathbf{x}_k^i]}, \xi_{q \to f_n}^{j,[\mathbf{x}_k^i]}). \tag{3.24}$$

The message passed from the $\mathbf{x}_q^j$-aligned sub-factor-graph, $\mathbf{I}(\mathbf{x}_q^j)$, can be given by

$$\mathbf{I}(\mathbf{x}_q^j) \propto \mathcal{CN}(\boldsymbol{\mu}_q^j, \boldsymbol{\xi}_q^j), \tag{3.25}$$

with mean $\boldsymbol{\mu}_q^j = [\mu_{q,1}^j, \ldots, \mu_{q,N}^j]$ and variance $\boldsymbol{\xi}_q^j = [\xi_{q,1}^j, \ldots, \xi_{q,N}^j]$. Then, $I_n(\mathbf{x}_q^j)$ can be given by

$$I_n(\mathbf{x}_q^j) \propto \mathcal{CN}(\mu_{q,n}^j, \xi_{q,n}^j). \tag{3.26}$$

In the following, the message passing operations from VN to FN and FN to VN are introduced respectively.

3.4.2.1 Message Passing from VN to FN

On the VN side, all but the k-th VNs need to be updated with the assistance of the corresponding sub-factor-graphs. Then, according to (3.16) and (3.26), the mean $\mu_{q \to f_n}^{j,[\mathbf{x}_k^i]}$

and the variance $\xi_{q \to f_n}^{j,[\mathbf{x}_k^i]}$ of message $I_{q \to f_n}^{j,[\mathbf{x}_k^i]}$ in the i_{iter}-th iteration can be respectively given by

$$\mu_{q \to f_n}^{j,[\mathbf{x}_k^i]} = \mu_{q,n}^{j}, \tag{3.27}$$

and

$$\xi_{q \to f_n}^{j,[\mathbf{x}_k^i]} = \xi_{q,n}^{j}, \tag{3.28}$$

where $q \neq k$, $j \in V_q^{[\mathbf{x}_k^i]}$, and the mean $\mu_{q,n}^{j}$ and the variance $\xi_{q,n}^{j}$ are obtained from previous iteration.

3.4.2.2 Message Passing from FN to VN

On the FN side, according to (3.4), the signal $y_{k,n}^i$ is contributed by $x_{k,n}^i$, $z_{k,n}^i$ and IUI. Since the symbol estimate of each interfering user is projected to a Gaussian distribution, the mean $\mu_{f_n \to k}^{i,[\mathbf{x}_k^i]}$ and the variance $\xi_{f_n \to k}^{i,[\mathbf{x}_k^i]}$ can be respectively derived based on the messages calculated in the current iteration, i.e.,

$$
\begin{aligned}
\mu_{f_n \to k}^{i,[\mathbf{x}_k^i]} = \frac{1}{\rho_{k,k} h_{k,n}} &\left(y_{k,n}^i - \sum_{l,l \neq k} h_{l,n} \rho_{l,k} \mu_{l \to f_n}^{i,[\mathbf{x}_k^i]} \right. \\
&\left. - \sum_{l<k} h_{l,n} v_{l,k} \mu_{l \to f_n}^{i+1,[\mathbf{x}_k^i]} - \sum_{l>k} h_{l,n} v_{l,k} \mu_{l \to f_n}^{i-1,[\mathbf{x}_k^i]} \right),
\end{aligned}
\tag{3.29}
$$

and

$$
\begin{aligned}
\xi_{f_n \to k}^{i,[\mathbf{x}_k^i]} = \frac{1}{|\rho_{k,k} h_{k,n}|^2} &\left(N_0 + \sum_{l,l \neq k} |h_{l,n} \rho_{l,k}|^2 \xi_{l \to f_n}^{i,[\mathbf{x}_k^i]} \right. \\
&\left. + \sum_{l<k} |h_{l,n} v_{l,k}|^2 \xi_{l \to f_n}^{i+1,[\mathbf{x}_k^i]} + \sum_{l>k} |h_{l,n} v_{l,k}|^2 \xi_{l \to f_n}^{i-1,[\mathbf{x}_k^i]} \right).
\end{aligned}
\tag{3.30}
$$

Define the vector $\mathbf{a}_k$ as a possible codeword of the k-th user and denote $a_{k,n}$ as its the n-th element. The posterior probability $p\left(\mathbf{x}_k^i = \mathbf{a}_k | \mathbf{Y}\right)$ can be calculated by

$$p\left(\mathbf{x}_k^i = \mathbf{a}_k | \mathbf{Y}\right) \propto I_{\Delta_k^i \to k}(\mathbf{a}_k) \prod_{n=1}^{N} I_{f_n \to k}^{i,[\mathbf{x}_k^i]}, \tag{3.31}$$

where $I_{\Delta_k^i \to k}(\mathbf{a}_k)$ denotes the prior probability for $\mathbf{x}_k^i = \mathbf{a}_k$ and $I_{f_n \to k}^{i,[\mathbf{x}_k^i]}$ can be given by

$$I_{f_n \to k}^{i,[\mathbf{x}_k^i]} \propto \frac{1}{\pi \xi_{f_n \to k}^{i,[\mathbf{x}_k^i]}} \exp\left(-\left| a_{k,n} - \mu_{f_n \to k}^{i,[\mathbf{x}_k^i]} \right|^2 / \xi_{f_n \to k}^{i,[\mathbf{x}_k^i]} \right). \tag{3.32}$$

Algorithm 3.1 Proposed 3D-EPA Detector

Input: spreading squence $\mathbf{w}_k$, modulation type, channel coefficient $h_{k,n}$, noise variance N_0, iteration number N_{iter}

Output: posterior LLRs $\Lambda\left(\mathbf{x}_k^i\right)$

  //**Initialization:**
  $\mu_{k,n}^i = 0, \xi_{k,n}^i = 1000, \forall k, i$
  //**Iteration:**
  for $i_{\text{iter}} = 1 : N_{\text{iter}}$ **do**
    for $i = 1 : L$ **do**
      for $k = 1 : K$ **do**
        //**VN update:**
        Compute $\mu_{q \to f_n}^{j,[\mathbf{x}_k^i]}$ and $\xi_{q \to f_n}^{j,[\mathbf{x}_k^i]}$, $\forall n$ via (3.27)–(3.28) for $q \neq k, \forall j \in V_q^{[\mathbf{x}_k^i]}$.
        //**FN update:**
        Compute $\mu_{f_n \to k}^{i,[\mathbf{x}_k^i]}$ and $\xi_{f_n \to k}^{i,[\mathbf{x}_k^i]}$, $\forall n$ via (3.29)–(3.30).
        Compute $p\left(\mathbf{x}_k^i | \mathbf{Y}\right)$ via (3.31).
        Compute $\mu_{k,n}^i$ and $\xi_{k,n}^i$, $\forall n$ via (3.33)–(3.34).
      end for
    end for
  end for
  // **LLR Calculation:**
  Compute posteriori LLRs $\Lambda\left(\mathbf{x}_k^i\right)$ via (3.9).

Further, we denote $\mathcal{X}_k$ as the codebook of the k-th user which is the collection of all its codewords. The cardinality $|\mathcal{X}_k|$ can be given as $|\mathcal{X}_k| = M$ and $\log_2 M$ represents the modulation order. Taking $p(\mathbf{x}_k^i)$ with equal probability into account, we have $I_{\Delta_k^i \to k}(\mathbf{a}_k) = p(\mathbf{x}_k^i = \mathbf{a}_k) = 1/M$. Then, the terms $\mu_{k,n}^i$ and $\xi_{k,n}^i$ can be respectively given by

$$\mu_{k,n}^i = \sum_{\mathbf{a}_k \in \mathcal{X}_k} p\left(\mathbf{x}_k^i = \mathbf{a}_k \mid \mathbf{Y}\right) a_{k,n}, \tag{3.33}$$

and

$$\xi_{k,n}^i = \sum_{\mathbf{a}_k \in \mathcal{X}_k} p\left(\mathbf{x}_k^i = \mathbf{a}_k \mid \mathbf{Y}\right) \left|a_{k,n} - \mu_{k,n}^i\right|^2. \tag{3.34}$$

Note that, similar to $\mu_{f_n \to k}^{i,[\mathbf{x}_k^i]}$ and $\xi_{f_n \to k}^{i,[\mathbf{x}_k^i]}$, only $p\left(\mathbf{x}_k^i | \mathbf{Y}\right)$, $\mu_{k,n}^i$, and $\xi_{k,n}^i$ are estimated in the $\mathbf{x}_k^i$-aligned sub-factor-graph. Whereas, the symbols $\mathbf{x}_q^j$ interfering with $\mathbf{x}_k^i$ are not estimated here but passed from the $\mathbf{x}_q^j$-aligned sub-factor-graphs to reduce the complexity overhead.

The frame-level iteration in the 3D-EPA is performed from the head to the tail of the entire frame in a zigzag, as shown in Fig. 3.2a. This is because the symbols in the head of a frame have lower IUI, especially for those symbols with smaller time delays. When the iteration is over, the posterior LLRs $\Lambda\left(\mathbf{x}_k^i\right)$ are calculated by (3.9). Finally, the proposed 3D-EPA is summarized in Algorithm 1. We note that this algorithm can also be deployed in systems with high order modulation.

3.4.3 Extension to Multiple Antenna Systems

In this subsection, we extend the 3D-EPA to multiple antenna systems. First, we consider a multiple antenna system with N_r receiving antennas. The receiving signal of the n_r-th path can be given by

$$y_n^{n_r}(t) = \sum_{k=1}^{K} h_{k,n}^{n_r} x_{k,n}(t - \tau_k) + z_n^{n_r}(t), \qquad (3.35)$$

where $h_{k,n}^{n_r}$ represents the channel coefficient of the k-th user on the n-th resource element and the n_r-th receiving antenna, and $z_n^{n_r}(t)$ represents the AWGN on $y_n^{n_r}(t)$. Then, the discrete symbol $y_{k,n}^{i,n_r}$ is generated by passing a group of matched filters and downsampling successively, which can be given by

$$
\begin{aligned}
y_{k,n}^i &= \Big[y_n(t) * MF(\tau_k) \Big]_{t=(i-1)T+\tau_k}^{iT+\tau_k} \\
&= h_{k,n}^{n_r} x_{k,n}^i + \sum_{m \neq k} h_{m,n}^{n_r} \rho_{m,k} x_{m,n}^i \\
&\quad + \sum_{m<k} h_{m,n}^{n_r} v_{m,k} x_{m,n}^{i+1} + \sum_{m>k} h_{m,n}^{n_r} v_{m,k} x_{m,n}^{i-1} + z_{k,n}^{i,n_r},
\end{aligned}
\qquad (3.36)
$$

where $z_{k,n}^{i,n_r}$ is the AWGN on $y_{k,n}^{i,n_r}$.

To fully characterize this multi-antenna system in the 3D factor-graph, the FNs in each sub-factor-graph have to be extended into N_r groups, which is shown in Fig. 3.4. Then, the 3D-EPA algorithm for multi-antenna systems can be obtained, by modifying $\mu_{f_n \to k}^{j,[\mathbf{x}_k^i]}$, $\xi_{f_n \to k}^{j,[\mathbf{x}_k^i]}$, $\mu_{q \to f_n}^{j,[\mathbf{x}_k^i]}$ and $\xi_{q \to f_n}^{j,[\mathbf{x}_k^i]}$ in the previous subsection to $\mu_{f_n^{n_r} \to k}^{j,[\mathbf{x}_k^i]}$, $\xi_{f_n^{n_r} \to k}^{j,[\mathbf{x}_k^i]}$, $\mu_{q \to f_n^{n_r}}^{j,[\mathbf{x}_k^i]}$ and $\xi_{q \to f_n^{n_r}}^{j,[\mathbf{x}_k^i]}$, respectively, and modifying (3.31) to

$$p\left(\mathbf{x}_k^i = \mathbf{a}_k | \mathbf{Y}\right) \propto I_{\Delta_k^i \to k}(\mathbf{a}_k) \prod_{n_r=1}^{N_r} \prod_{n=1}^{N} I_{f_n^{n_r} \to k}^{i,[\mathbf{x}_k^i]}. \qquad (3.37)$$

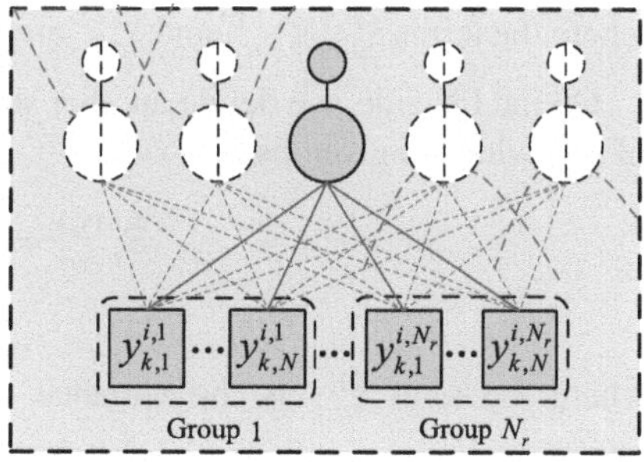

Fig. 3.4 Sub-factor-graph in 3D factor-graph for multi-antenna systems with N_r receiving antennas

Finally, the 3D-EPA for multiple antenna systems can be derived, which is similar to Algorithm 1.

3.4.4 State Evolution Analysis

State evolution is a statistical analysis method to deduce the message distributions [27, 42] which is often applied to analyze multi-user detection algorithms such as AMP [43], EPA [44], etc., to predict system performance. In this subsection, we develop the state evolution based on the 3D-EPA for asynchronous code-domain NOMA.

Consider a multiple antenna system with N_r receiving antennas. On the VN side, we introduce a state evolution parameter $\Xi_{n,i_{\text{iter}}}^{[\mathbf{x}_k^i]}$, which is defined as

$$\Xi_{n,i_{\text{iter}}}^{[\mathbf{x}_k^i]} = \frac{\sum_{q=1}^{K \setminus k} \left(|\rho_{q,k}|^2 \xi_{q \to f_n^{nr}}^{i,[\mathbf{x}_k^i]} + |\nu_{q,k}|^2 \xi_{q \to f_n^{nr}}^{i-1,[\mathbf{x}_k^i]} + |\nu_{q,k}|^2 \xi_{q \to f_n^{nr}}^{i+1,[\mathbf{x}_k^i]} \right)}{K - 1}, \tag{3.38}$$

where the terms $\xi_{q \to f_n^{nr}}^{i,[\mathbf{x}_k^i]}, \xi_{q \to f_n^{nr}}^{i-1,[\mathbf{x}_k^i]}$, and $\xi_{q \to f_n^{nr}}^{i+1,[\mathbf{x}_k^i]}$ are all generated in the $(i_{\text{iter}} - 1)$-th iteration. According to (3.28), $\xi_{q \to f_n^{nr}}^{i,[\mathbf{x}_k^i]}, q \neq k$, and $\xi_{q \to f_n^{nr}}^{i\pm1,[\mathbf{x}_k^i]}, q \neq k$, can be respectively substituted by

$$\xi_{q \to f_n^{nr}}^{i,[\mathbf{x}_k^i]} = \xi_{q,n}^i, \tag{3.39}$$

and

$$\xi_{q \to f_n^{nr}}^{i\pm1,[\mathbf{x}_k^i]} = \xi_{q,n}^{i\pm1}. \tag{3.40}$$

Then, by substituting (3.39) and (3.40) into (3.38), $\Xi_{n,i_{\text{iter}}}^{[\mathbf{x}_k^i]}$ can be approximately given by

$$\Xi_{n,i_{\text{iter}}}^{[\mathbf{x}_k^i]} \approx \frac{1}{K - 1} \sum_{q=1}^{K \setminus k} \left(|\rho_{q,k}|^2 \xi_{q,n}^i + |\nu_{q,k}|^2 \xi_{q,n}^{i-1} + |\nu_{q,k}|^2 \xi_{q,n}^{i+1} \right), \tag{3.41}$$

where the terms $\xi_{q,n}^i, \xi_{q,n}^{i-1}$, and $\xi_{q,n}^{i+1}$ are all generated in the $(i_{\text{iter}} - 1)$-th iteration.

On the FN side, we define another state evolution parameter $\hat{\Xi}_{n,i_{\text{iter}}}^{i,[\mathbf{x}_k^i]}$ as the output of FN, which is given by

$$\hat{\Xi}_{n,i_{\text{iter}}}^{i,[\mathbf{x}_k^i]} = \frac{1}{K} \sum_{q=1}^{K} \hat{\xi}_{q,n}^i, \tag{3.42}$$

where the term $\hat{\xi}_{q,n}^i$ is the variance of $p\left(\mathbf{x}_k^i = \mathbf{a}_k | \mathbf{Y}\right)$ in (3.37). According to operational rule of exponential distribution, $\hat{\xi}_{q,n}^i$ can be rewritten as

$$\hat{\xi}_{q,n}^{i} = \Big(\sum_{n_r=1}^{N_r} \frac{1}{\hat{\xi}_{f_n^{n_r} \to q}^{i,[\mathbf{x}_q^i]}}\Big)^{-1}.$$

(3.43)

Then, substituting (3.30) into (3.43), (3.44) is obtained and shown at the top of the next page, where $(*)$ can be approximated as

$$\hat{\xi}_{q,n}^{i} = \left(\sum_{n=1}^{N} \frac{|\rho_{k,k} h_{k,n}|^2}{\underbrace{N_0 + \sum_{l,l \neq k} |h_{l,n}\rho_{l,k}|^2 \xi_{l \to f_n^{n_r}}^{i,[\mathbf{x}_k^i]} + \sum_{l<k} |h_{l,n} v_{l,k}|^2 \xi_{l \to f_n^{n_r}}^{i+1,[\mathbf{x}_k^i]} + \sum_{l>k} |h_{l,n} v_{l,k}|^2 \xi_{l \to f_n^{n_r}}^{i-1,[\mathbf{x}_k^i]}}_{(*)}}\right)^{-1}.$$

(3.44)

$$(*) \approx \sum_{l=1}^{K \backslash k} \Big(|h_{l,n}\rho_{l,k}|^2 \xi_{l \to f_n^{n_r}}^{i,[\mathbf{x}_k^i]} + |h_{l,n} v_{l,k}|^2 \xi_{l \to f_n^{n_r}}^{i+1,[\mathbf{x}_k^i]}$$
$$+ |h_{l,n} v_{l,k}|^2 \xi_{l \to f_n^{n_r}}^{i-1,[\mathbf{x}_k^i]}\Big).$$

(3.45)

Then, by substituting (3.38) and (3.44) into (3.42), $\hat{\Xi}_{n,i_{\text{iter}}}^{i,[\mathbf{x}_k^i]}$ can be finally rewritten as

$$\hat{\Xi}_{n,i_{\text{iter}}}^{i,[\mathbf{x}_k^i]} = \frac{N_0 + (K-1)\Xi_{n,i_{\text{iter}}}^{[\mathbf{x}_k^i]}}{N_r}.$$

(3.46)

By iterations of (3.41) and (3.46), the iterative performance of the 3D-EPA can be predicted.

3.5 Simulation Results and Complexity Analysis

In this section, the BER performance of asynchronous code-domain NOMA with the 3D-EPA is evaluated with uniformly distributed time delay. We also compare its BER performance with the synchronous counterpart and other asynchronous multi-user detection algorithms. Besides, the convergence, state evolution, and complexity of the proposed algorithm are analyzed. Simulation parameters are summarized in Table 3.1.

3.5.1 Simulation Results

Figure 3.5 compares the BER performance of the proposed 3D-EPA under different frame lengths and overloadings in the AWGN channel. In the case with $\eta = 150\%$,

Table 3.1 Simulation settings

Parameter	Assumption
Modulation	BPSK
Spreading sequence $\mathbf{w}_k$	WBE-based codes
Waveform $s(t)$	Rooted raised cosine
Roll-off factor α	0.6
Delay τ_k	Uniform distribution over $[0, T]$
Overloading η	150%, 200%, 300%
Spreading factor N	4, 6
Channel	AWGN, Rayleigh fading
Frame length L	4, 8, 16
Iterations N_{iter}	5

Fig. 3.5 Comparisons on BER performance under different overloading η and frame length L

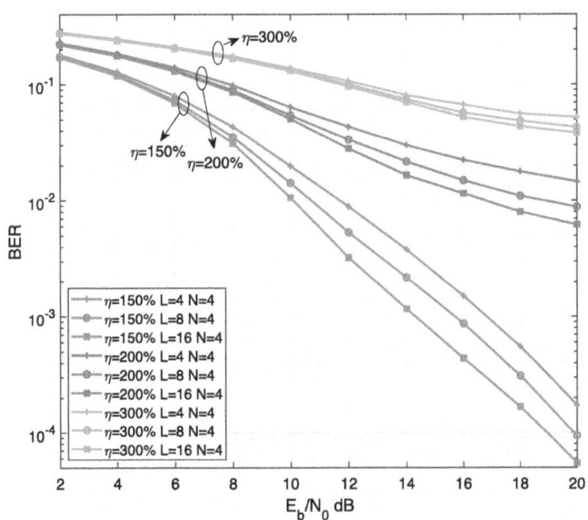

the BER performance of the 3D-EPA can reach 10^{-4} without error floor. Meanwhile, the BER performance can also gradually improve with the increasing frame length L, as yielded in the spatial coupling theory [45]. This means that a modest increase in frame length can yield a trade-off between complexity and performance. Besides, under higher overloadings, such as $\eta = 200\%$ and $\eta = 300\%$, the BER performance deteriorates and the error floor occurs. This is because the 3D-EPA accumulates serious errors due to the Gaussian approximation of the probability distribution.

Figure 3.6 compares the performance between the asynchronous NOMA with the 3D-EPA and synchronous code-domain NOMA with EPA [26] under different overloadings in the AWGN channel. Under $\eta = 200\%$ and $\eta = 300\%$, the proposed 3D-EPA shows a significant gain compared with synchronous code-domain NOMA with the EPA. This is because the interference of symbols in adjacent time intervals, e.g.,

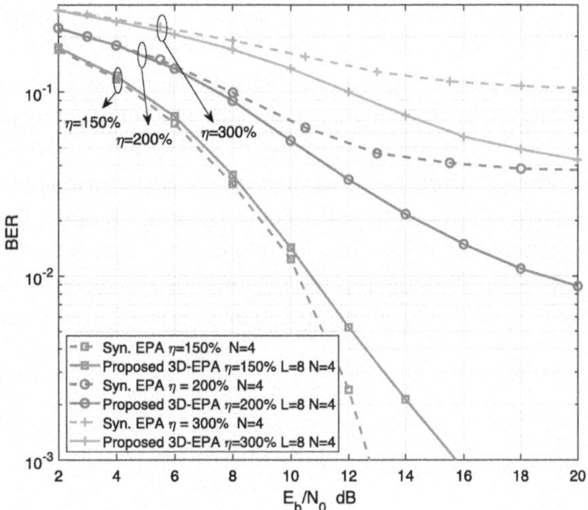

Fig. 3.6 BER performance of asynchronous and synchronous NOMA with different detection schemes

$\rho_{m,k}x_{m,n}^{i}$ and $v_{m,k}x_{m,n}^{i+1}$, can partially cancel out on $y_{k,n}^{i}$, $k > m$ in the asynchronous case. However, under lower overloading, i.e., $\eta = 150\%$, the increased number of nodes in the 3D factor-graph is the dominant factor, which results in worse BER performance than synchronous NOMA with the EPA. The above results indicate that asynchronous code-domain NOMA with 3D-EPA outperforms its synchronous counterpart, especially under high overloadings.

Figure 3.7 compares the BER performance of the 3D-EPA under different spreading factor in the AWGN channel. Under the same overloading, the BER performance

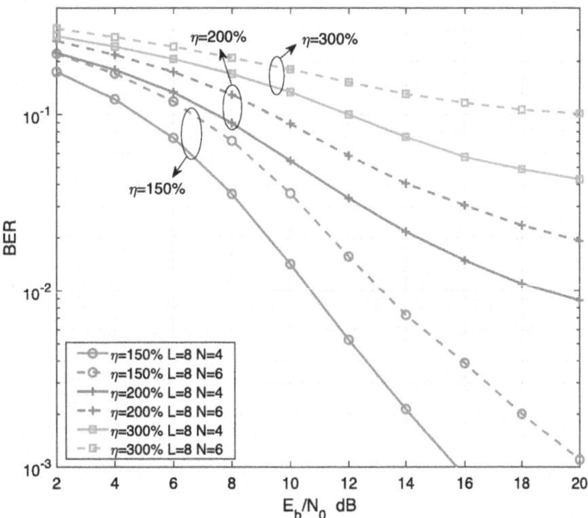

Fig. 3.7 BER performance with different spreading factor N

of the case with a small spreading factor N outperforms that with a large one. The reason is that the coding gain brought by the increasing N cannot compensate for the performance deterioration caused by the more serious asynchronous IUI. This indicates that the BER performance gain cannot be achieved by increasing the spreading factor.

The 3D-EPA can also be extended to multi-antenna systems with simple modifications. Figure 3.8 compares the BER performance of the 3D-EPA with different number of receiving antennas in Rayleigh fading channel. The results show that a significant BER gain can be obtained by increasing N_r regardless of the size of N.

Figure 3.9 compares the performance of the proposed 3D-EPA with MMSE and MMSE-SIC [33] under different overloadings. It can be observed that the BER

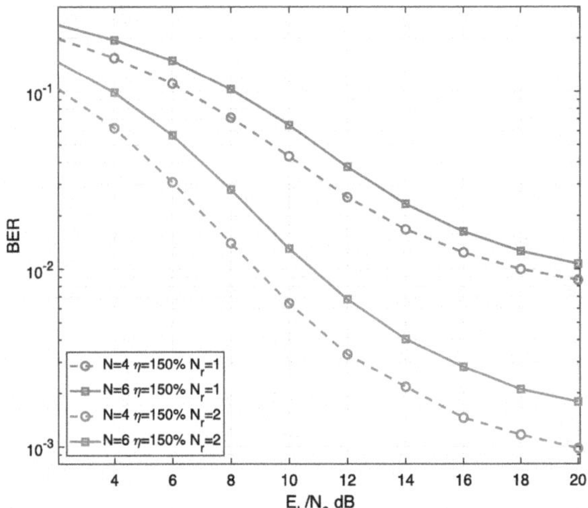

Fig. 3.8 BER performance with different receiving antenna N_r in Rayleigh channel

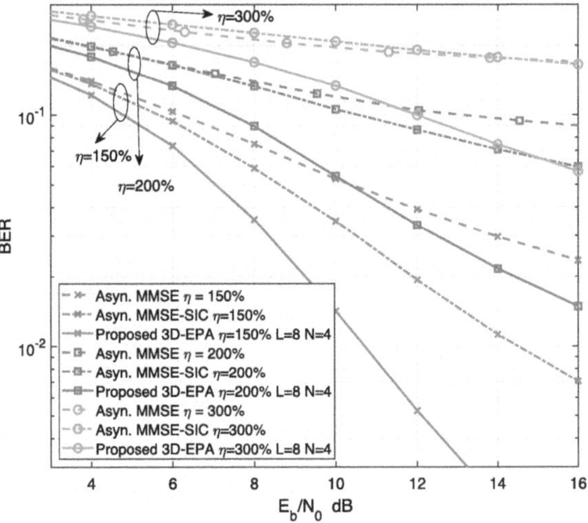

Fig. 3.9 BER performance comparison of the 3D-EPA, MMSE and MMSE-SIC [33] in the AWGN channel

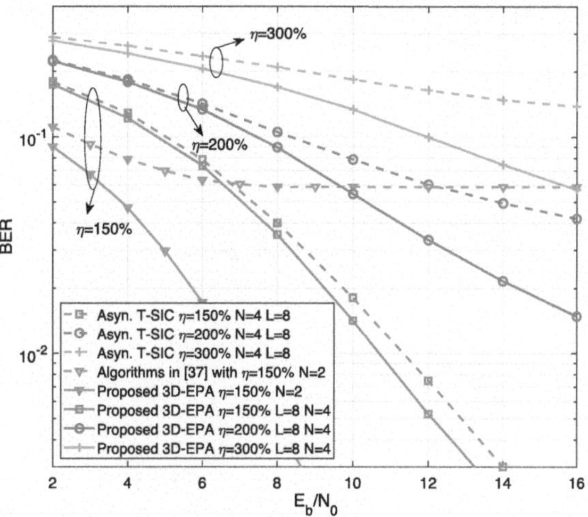

Fig. 3.10 BER performance comparison of the 3D-EPA, T-SIC [35] and the algorithm in [36] under the AWGN channel

performance of the 3D-EPA significantly outperforms that of the other two algorithms. Even though the MMSE-SIC outperforms the MMSE by using the SIC, it is still worse than the 3D-EPA due to the error propagation and the inability to take advantage of the interference structure. In addition, Fig. 3.10 compares the BER performance of the proposed 3D-EPA, the triangle SIC (T-SIC) [35] and the algorithm in [36] for code-domain NOMA. Compared with the T-SIC, the proposed 3D-EPA can achieve about 0.6dB gain at the BER of 10^{-2} under $\eta = 150\%$, and the BER gain under other overloadings is also significant. The reason is similar to the conclusion of Fig. 3.9. Compared with the algorithm in [36], the 3D-EPA also has significant BER performance gain. This is because the ESE used in [36] can be regarded as a simplification of the EPA [46], which requires the assistance of channel coding to refine detection performance. Moreover, this algorithm does not consider symbol spreading, so spreading gain can not be obtained.

Figure 3.11 analyzes the convergence of the 3D-EPA under different overloadings and E_b/N_0 in the AWGN channel. It can be observed that the 3D-EPA under $\eta = 150\%$ normally converges after 3 iterations. By contrast, under the higher overloading, the 3D-EPA requires 4 iterations to converge, no matter $E_b/N_0 = 10$dB or 15 dB, which is attributed to the increased IUI under high overloadings.

In addition, Fig. 3.12 analyzes the state evolution of the proposed 3D-EPA in asynchronous transmissions. It shows that the 3D-EPA matches the theoretical curve at the low SNR region. However, the two curves gradually separate at the high SNR region due to the limited system scale [27, 44].

3.5.2 Complexity Analysis

In this subsection, the complexity of the 3D-EPA is analyzed. Table 3.2 sums up the complexity of the 3D-EPA and several other multi-user detection algorithms.

Fig. 3.11 Convergence of the 3D-EPA with different η and E_b/N_0

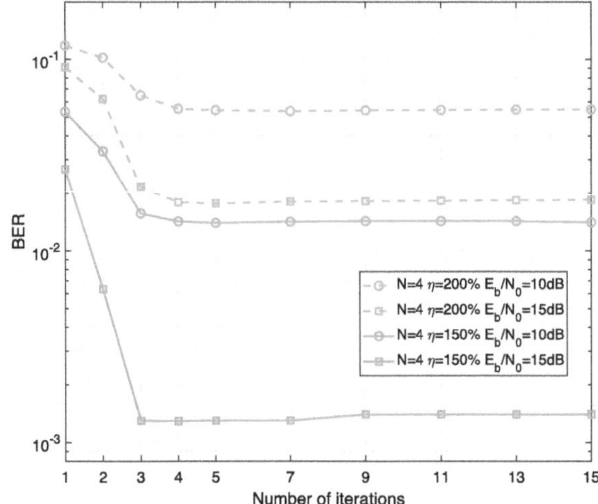

Fig. 3.12 State evolution analysis for the 3D-EPA

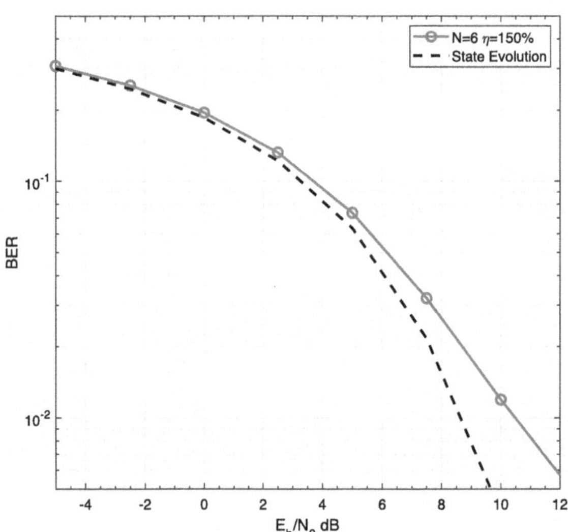

Table 3.2 Complexity comparison

Receiver type	Complexity order
3D-EPA (asyn.)	$\mathcal{O}(N_{\text{iter}} L K N M)$
EPA (syn.) [26]	$\mathcal{O}(N_{\text{iter}} L K N M)$
MMSE (asyn.)	$\mathcal{O}(L K N^3)$
MMSE-based T-SIC (asyn.) [35]	$\mathcal{O}(N_{\text{iter}} L K^2 N^3)$

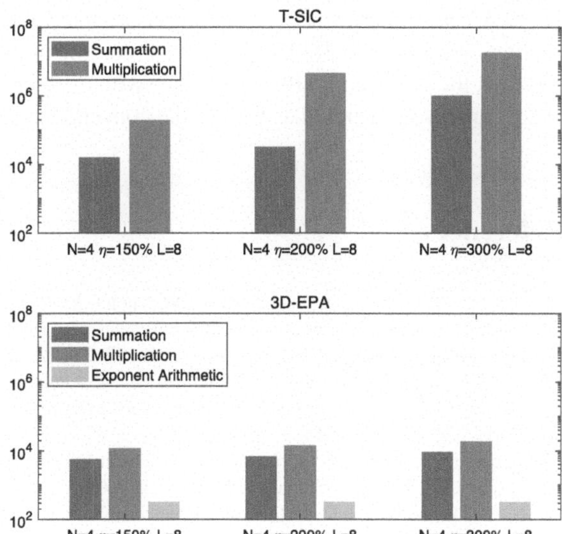

Fig. 3.13 Comparisons on the amounts of different operation types between 3D-EPA and T-SIC

Note that only the dominant terms are included in the order of complexity. It can be observed that the 3D-EPA has the same linear complexity as traditional EPA. This is because in the proposed algorithm, each sub-factor-graph only performs the expectation propagation for a single symbol, and finally, multiple sub-factor-graphs work together to complete the estimation of all symbols in the entire frame. In addition, we also compared the 3D-EPA with another algorithm designed for asynchronous transmissions. Figure 3.13 compares the complexity of the 3D-EPA with T-SIC in terms of summation, multiplication, and exponential arithmetic. It can be seen that the 3D-EPA outperforms the T-SIC under different overloadings, especially under high overloadings. Combined with the conclusion drawn in Fig. 3.10, this result indicates that the 3D-EPA can achieve better performance as well as lower complexity.

3.6 Conclusions

This chapter proposes a multi-user detection algorithm for asynchronous code-domain NOMA. By introducing the time delay dimension in the traditional factor-graph, the 3D factor-graph model is constructed with two message passing mechanisms to characterize the interference structure of asynchronous code-domain NOMA. Then, the low-complexity 3D-EPA is developed based on this model to modify the interference approximation. Furthermore, we also extend the 3D-EPA to multi-antenna systems and derive the state evolution to analyze the theoretical BER performance. Finally, based on the link-level simulations, we conclude the following remarks: first, the BER performance of asynchronous code-domain NOMA with the

3D-EPA can outperform its synchronous counterpart, especially under high over-loadings. Second, the 3D-EPA has linear complexity similar to traditional EPA. In addition, the achieved BER performance with different frame lengths and spreading factors are also illustrated as well as the convergence performance of the 3D-EPA.

References

1. S. Chen, Y.-C. Liang, S. Sun, S. Kang, W. Cheng, M. Peng, Vision, reianuirements, and technology trend of 6G: how to tackle the challenges of system coverage, capacity, user data-rate and movement speed. IEEE Wirel. Commun. **27**(2), 218–228 (2020)
2. A. Al-Fuqaha, M. Guizani, M. Mohammadi, M. Aledhari, M. Ayyash, Internet of things: a survey on enabling technologies, protocols, and applications. IEEE Commun. Surveys Tutorials **17**(4), 2347–2376 (2015)
3. L. Chettri, R. Bera, A comprehensive survey on Internet of things (IoT) toward 5G wireless systems. IEEE Internet Things J. **7**(1), 16–32 (2020)
4. N. Ye, J. An, J. Yu, Deep-learning-enhanced NOMA transceiver design for massive MTC: challenges, state of the art, and future directions. IEEE Wirel. Commun. **28**(4), 66–73 (2021)
5. Z. Ding, Y. Liu, J. Choi, Q. Sun, M. Elkashlan, I. Chih-Lin, H.V. Poor, Application of non-orthogonal multiple access in LTE and 5G networks. IEEE Commun. Mag. **55**(2), 185–191 (2017)
6. H. Yu, N. Ye, A. Wang, Non-orthogonal wireless backhaul design for cell-free massive MIMO: an integrated computation and communication approach. IEEE Wirel. Commun. Lett. **10**, 281–285 (2021)
7. Z. Ding, X. Lei, G.K. Karagiannidis, R. Schober, J. Yuan, V.K. Bhargava, A survey on non-orthogonal multiple access for 5G networks: research challenges and future trends. IEEE J. Sel. Areas Commun. **35**(10), 2181–2195 (2017)
8. L. Dai, B. Wang, Y. Yuan, S. Han, I. Chih-lin, Z. Wang, Non-orthogonal multiple access for 5G: solutions, challenges, opportunities, and future research trends. IEEE Commun. Mag. **53**(9), 74–81 (2015)
9. J. Pan, N. Ye, H. Yu, T. Hong, S. Al-Rubaye, S. Mumtaz, A. Al-Dulaimi, I. Chih-Lin, Ai-driven blind signature classification for IoT connectivity: a deep learning approach, *IEEE Transactions on Wireless Communications*, pp. 1–1 (2022)
10. L. Dai, B. Wang, Z. Ding, Z. Wang, S. Chen, L. Hanzo, A survey of non-orthogonal multiple access for 5G. IEEE Commun. Surveys Tutorials **20**(3), 2294–2323 (2018)
11. N. Ye, X. Li, J. Pan, W. Liu, X. Hou, Beam aggregation-based mmwave MIMO-NOMA: an AI-enhanced approach. IEEE Trans. Veh. Technol. **70**(3), 2337–2348 (2021)
12. W. Saad, M. Bennis, M. Chen, A vision of 6G wireless systems: applications, trends, technologies, and open research problems. IEEE Network **34**(3), 134–142 (2019)
13. N. Ye, J. Yu, A. Wang, R. Zhang, Help from space: grant-free massive access for satellite-based IoT in the 6G era. Digital Commun. Networks **8**, 215–224 (2022)
14. M.B. Shahab, R. Abbas, M. Shirvanimoghaddam, S.J. Johnson, Grant-free non-orthogonal multiple access for IoT: a survey. IEEE Commun. Surveys Tutorials **22**(3), 1805–1838 (2020)
15. O. Kodheli et al., Satellite communications in the new space era: a survey and future challenges. IEEE Commun. Surveys Tutorials **23**(1), 70–109 (2020)
16. S.R. Islam, N. Avazov, O.A. Dobre, K.-S. Kwak, Power-domain non-orthogonal multiple access (NOMA) in 5G systems: Potentials and challenges. IEEE Commun. Surveys Tutorials **19**(2), 721–742 (2016)
17. H. Nikopour, H. Baligh, Sparse code multiple access, in *Proceedings IEEE 24th Annual International Symposium on Personal, Indoor, and Mobile Radio Communications (PIMRC)*, pp. 332–336, IEEE (2013)

18. Ericsson, "Signature design for NoMA," r1-1806241, ran wg1 meeting no. 93, (2018)
19. N. Ye, X. Li, H. Yu, L. Zhao, W. Liu, X. Hou, DeepNOMA: a unified framework for NOMA using deep multi-task learning. IEEE Trans. Wirel. Commun. **19**(4), 2208–2225 (2020)
20. 3GPP TR 38.812, Study on non-orthogonal multiple access (NOMA) for NR (2018)
21. S. Moshavi, Multi-user detection for DS-CDMA communications. IEEE Commun. Mag. **34**(10), 124–136 (1996)
22. R. Hoshyar, F.P. Wathan, R. Tafazolli, Novel low-density signature for synchronous CDMA systems over AWGN channel. IEEE Trans. Signal Process. **56**(4), 1616–1626 (2008)
23. W. Yuan, N. Wu, Q. Guo, D. Ng, J. Yuan, L. Hanzo, Iterative joint channel estimation, user activity tracking, and data detection for FTN-NOMA systems supporting random access. IEEE Trans. Commun. **68**(5), 2963–2977 (2020)
24. S. Rangan, P. Schniter, A.K. Fletcher, Vector approximate message passing. IEEE Trans. Inform. Theory **65**(10), 6664–6684 (2019)
25. J. Ma, L. Liu, X. Yuan, L. Ping, On orthogonal AMP in coded linear vector systems. IEEE Trans. Wirel. Commun. **18**(12), 5658–5672 (2019)
26. X. Meng, Y. Wu, Y. Chen, M. Cheng, Low complexity receiver for uplink SCMA system via expectation propagation, in *2017 IEEE Wireless Communications and Networking Conference (WCNC)*, pp. 1–5 (2017)
27. P. Wang, L. Liu, S. Zhou, G. Peng, S. Yin, S. Wei, Near-optimal MIMO-SCMA uplink detection with low-complexity expectation propagation. IEEE Trans. Wirel. Commun. **19**(2), 1025–1037 (2019)
28. J. Céspedes, P.M. Olmos, M. Sánchez-Fernández, F. Perez-Cruz, Expectation propagation detection for high-order high-dimensional MIMO systems. IEEE Trans. Commun. **62**(8), 2840–2849 (2014)
29. D. Zhang, L.L. Mendes, M. Matthè, I.S. Gaspar, N. Michailow, G.P. Fettweis, Expectation propagation for near-optimum detection of MIMO-GFDM signals. IEEE Trans. Wirel. Commun. **15**(2), 1045–1062 (2016)
30. Y. Zhang, Z. Yuan, Q. Guo, Z. Wang, J. Xi, Y. Li, Bayesian receiver design for grant-free NOMA with message passing based structured signal estimation. IEEE Trans. Veh. Technol. **69**(8), 8643–8656 (2020)
31. Y. Ge, Q. Deng, P.C. Ching, Z. Ding, OTFS signaling for uplink NOMA of heterogeneous mobility users. IEEE Trans. Commun. **69**(5), 3147–3161 (2021)
32. W. Yuan, N. Wu, A. Zhang, X. Huang, Y. Li, L. Hanzo, Iterative receiver design for FTN signaling aided sparse code multiple access. IEEE Trans. Wirel. Commun. **19**(2), 915–928 (2020)
33. M. Ganji, X. Zou, H. Jafarkhani, Asynchronous transmission for multiple access channels: Rate-region analysis and system design for uplink NOMA. IEEE Trans. Wirel. Commun. **20**(7), 4364–4378 (2021)
34. H. Lee, I. Jung, J. Heo, D. Hong, Exploiting intentional time-domain offset in downlink multicarrier NOMA systems. IEEE Wirel. Commun. Lett. **10**(7), 1577–1580 (2021)
35. H. Haci, H. Zhu, J. Wang, Performance of non-orthogonal multiple access with a novel asynchronous interference cancellation technique. IEEE Trans. Wirel. Commun. **65**(3), 1319–1335 (2017)
36. J. Liu, Y. Li, G. Song, Y. Sun, Detection and analysis of symbol-asynchronous uplink NOMA with equal transmission power. IEEE Wirel. Commun. Lett. **8**(4), 1069–1072 (2019)
37. J. Wang, C. Jiang, L. Kuang, Iterative NOMA detection for multiple access in satellite high-mobility communications. IEEE J. Sel. Areas Commun. **40**(4), 1101–1113 (2022)
38. Q. Yu, H. Li, W. Meng, W. Xiang, Sparse code multiple access asynchronous uplink multiuser detection algorithm. IEEE Trans. Veh. Technol. **68**(6), 5557–5569 (2019)
39. S. Verdu, The capacity region of the symbol-asynchronous gaussian multiple-access channel. IEEE Trans. Inform. Theory **35**(4), 733–751 (1989)
40. M.I. Jordan, An introduction to probabilistic graphical models (2003)
41. T.P. Minka, Expectation propagation for approximate Bayesian inference, in *Proceedings 17th Conference on Uncertainty in Artificial Intelligence*, pp. 362–369 (2001)

42. F. Krzakala, M. Mézard, F. Sausset, Y. Sun, L. Zdeborová, Probabilistic reconstruction in compressed sensing: algorithms, phase diagrams, and threshold achieving matrices. J. Stat. Mech. Theory Exp. **2012**(08), P08009 (2012)
43. X. Meng, S. Wu, L. Kuang, J. Lu, Concise derivation of complex Bayesian approximate message passing via expectation propagation, arXiv preprint arXiv:1509.08658 (2015)
44. A. Kosasih, O. Setyawati, Rahmadwati, Low complexity multi-user MIMO detection for uplink SCMA system using expectation propagation algorithm. TELKOMNIKA Telecommun. Comput. Electron. Control **16**(1), 182–188 (2018)
45. C. Schlegel, D. Truhachev, Multiple access demodulation in the lifted signal graph with spatial coupling. IEEE Trans. Inform. Theory **59**(4), 2459–2470 (2013)
46. X. Meng, L. Zhang, C. Wang, L. Wang, Y. Wu, Y. Chen, W. Wang, Advanced NOMA receivers from a unified variational inference perspective. IEEE J. Sel. Areas Commun. **39**(4), 934–948 (2021)

Chapter 4
Efficient Coding: Quasi-Cyclic LDPC Transceiving System

In this chapter, we center around the QC-LDPC transceiving system with efficient coding. Section 4.1 introduces the background and motivation of QC-LDPC design. Section 4.2 introduces QC-LDPC codes and overlapped decoding scheme, and discusses the construction of the proposed QC-LDPC code optimized for overlap in detail. Section 4.3 reviews the Min-Sum Algorithm and Modified 2-bit MSA, and illustrates the simplification of combinatorial logic to realize the algorithm. Section 4.4 describes the system architecture. Section 4.5 presents the data stream and scheduling structure. Section 4.6 analyzes the simulation results and experimental results of our decoder. Section 4.7 concludes this chapter.

4.1 Introduction

Low-Density-Parity-Check (LDPC) codes are linear block codes that correct errors by adding parity bits, first proposed by R. Gallager in 1962 [1–3]. These codes are known to approach the channel capacity, with a code construction that allows them to work at a noise threshold close to the Shannon limit in memory-free symmetric channels [4, 5]. In other words, LDPC codes can be transmitted at a rate lower than the channel capacity while maintaining a low bit error rate through logical coding schemes [6]. Additionally, the sparsity of the PCM used in LDPC makes it feasible to implement LDPC codes on field-programmable gate arrays (FPGAs) on a large scale [7]. As a result, LDPC codes have been widely adopted in both wired and wireless standards such as IEEE 802.3an, IEEE 802.3ba, IEEE 802.11n, IEEE 802.16e, DVB-S2, CCSDS, and others [8].

Quasi-Cyclic LDPC (QC-LDPC) codes are a type of LDPC that can offer several advantages over other error-correcting codes, including low-complexity implementation, high error correction performance, flexibility, and wide usage in practical applications. They can be efficiently encoded using simple operations such as shifting and XOR [9]. Their regular structure also makes them easier to decode, and they

J. Li et al., *Key Technologies of High Frequency Wireless Communications*,
https://doi.org/10.1007/978-981-96-5894-7_4

can be designed to have a high code rate while still maintaining good error correction performance [10]. Additionally, they are also highly flexible, and can be designed to meet the requirements of different applications, including wireless communication systems, optical communication systems, and storage systems.CCSDS (8176, 7154) LDPC perfectly inherits the characteristics of QC-LDPC and maintains a high code rate, and has therefore been adopted for multiple situations [11].

The data processing speed of many specialized sensors, such as optical sensors, intelligent sensors, and biomaterial-based sensors, has been a pressing need for improvement [12–14]. With the intensification of the demand for communication speed in the sensor networks, the requirements for coding gain and throughput pertaining to channel coding in high-speed communication systems have correspondingly escalated [15]. While various decoding methods have been proposed, achieving high throughput with minimal hardware requirements and low power consumption remains a challenge. For instance, a LDPC decoder for 5G NR on FPGA proposed by Pourjabar achieves a maximum throughput of 2.2 Gbps at 10 iterations, but consumes 96 block RAMs and 225,191 look-up tables (LUT) [16]. Similarly, Sham's decoder for a (1944, 1620) LDPC codeachieves a throughput of 1.8 Gbps, requiring 100,000 registers, 65,000 LUTs, and 22 KB RAMs, which indicates low hardware utilization efficiency (HUE) [17]. And the decoder for ultra-long code with size of $149{,}504 \times 262{,}144$in [18] consumes 51,000 LUTs, 1000 DSPs, and 32 Mb RAMs, reaching a throughput of 108 mbps,but can work at a very low SNR.Nevertheless, designing high-performance LDPC decoders with minimal resource requirements and low power consumption remains a formidable challenge.

The conventional decoder designs of LDPC codes, which employ the variable node processing unit (VNU) and check node processing unit (CNU) sequentially in each iteration cycle, often result in low HUE and low throughput. Although data interleaving has been proposed to enhance HUE by allowing variable node processing (VNP) and check node processing (CNP) to work on separate block data simultaneously, it doubles the memory requirements [19]. Folding is another technique that combines different block data to achieve the ideal memory depth, but sacrifices the decoder's throughput [7]. To address these limitations, overlapped message passing has been proposed as an algorithm-level solution to overlap the CNU and VNU operations [20]. To maximize the overlap depth, which is the number of CNUs and VNUs concurrently involved in a single iteration, and enhance the throughput, a permutation vector-based LDPC code construction approach has been proposed. By using a partial parallel architecture, the proposed decoder allows column processing to commence after three row processing calculations instead of waiting for all row processing to complete, resulting in a one-third reduction in decoding time. In addition, a shift-register-based memory strategy has been employed to reduce the read/write latency.

The choice of decoding algorithm is a crucial factor affecting the performance of LDPC decoders. The MSA simplifies the multiplication operation in the decoding process to an addition operation, and approximates the summation process to finding the minimum value, thereby significantly reducing the algorithm's complexity [21]. Building on this, the Modified 2-bit MSA optimizes the storage structure and utilizes two bits to represent the information in the operation, which further reduces the resource overhead [22].

In this chapter, in response to the inability to perform deep overlap decoding for CCSDS (8176, 7154) LDPC codes, we propose a method for designing a QC-LDPC that is suitable for overlap decoding, while maintaining an equivalent submatrix structure and code length. We design a specialized partially parallel decoder based on the Modified 2-bit MSA for our proposed LDPC code, which exhibits a substantial coding gain of 5 dB at a BER of 10^{-6} compared to uncoded BPSK. Furthermore, when compared to the original CCSDS (8176, 7154) LDPC code with Modified 2-bit MSA decoding, our proposed LDPC maintains high performance with a marginal coding gain loss of less than 0.5 dB, while improving throughput and HUE.

4.2 Architecture of CCSDS-Like QC-LDPC Code for Overlap

4.2.1 QC-LDPC Code

QC-LDPC codes are a type of LDPC codes whose PCM can be decomposed into cyclic submatrices of equal size [23]. For an (n, k) LDPC code with a PCM of size $m \times n$ and submatrix dimension l, n and k represent code length before and after coding, respectively, and the PCM can be constructed using the base graph and shift factors $S_{m,n}$, where $0 \leq S_{m,n} \leq l - 1$. The 1s in the base graph are replaced by submatrices, and the 0s are replaced by zero matrices [24]. The origin submatrix can be either an identity matrix or a matrix derived from finite geometry, such as a double diagonal matrix. The shift factors specify the number of bits each submatrix should move to the right. For example, the cyclic submatrix $I(1)$ is obtained by shifting the origin double diagonal submatrix one bit to the right.

$$Q(1) = \begin{pmatrix} 0\ 1\ 1\ 0\ \cdots\ 0 \\ 0\ 0\ 1\ 1\ \cdots\ 0 \\ 0\ 0\ 0\ 1\ \cdots\ 0 \\ \vdots\ \vdots\ \vdots\ \vdots\ \ddots\ \vdots \\ 1\ 0\ 0\ 0\ \cdots\ 1 \\ 1\ 1\ 0\ 0\ \cdots\ 0 \end{pmatrix}_{l \times l}.$$

Specifically $Q(-1)$ denotes the zero matrix. And in this way, we can construct H, where m_a, n_b indicate the number of submatrices in horizontal and vertical distribution:

$$H = \begin{pmatrix} Q(S_{1,1})\ \ Q(S_{1,2})\ \cdots\ Q(S_{1,n_b}) \\ Q(S_{2,1})\ \ Q(S_{2,2})\ \cdots\ Q(S_{2,n_b}) \\ \vdots\ \ \ \ \ \ \ \ \vdots\ \ \ \ddots\ \ \ \ \vdots \\ Q(S_{m_a,1})\ Q(S_{m_a,2})\ \cdots\ Q(S_{m_a,n_b}) \end{pmatrix}_{m \times n}.$$

The PCM of CCSDS (8176, 7154) LDPC is presented in Fig. 4.1, whose block length is 8176 bits and the message length is 7154 bits. The PCM of CCSDS (8176, 7154) LDPC is composed of 32 submatrices in 2 rows and 16 columns. The size of submatrices is 511×511, and each submatrix satisfies the QC characteristic.

4.2.2 Overlapped Decoding Scheme

In traditional partially parallel decoder architectures, CNU and VNU are performed sequentially in each iterative cycle, as shown in Fig. 4.2a, resulting in low HUE and low throughput. To accelerate the decoding process, and keep low memory consumption at the same time, an overlapped decoding scheme is proposed [25]. An overlapped decoding scheme allows CNU and VNU to work independently at the same time, as shown in Fig. 4.2b, and in ideal situations, CNU and VNU can seamlessly overlap to achieve maximum throughput like Fig. 4.2c. In this chapter, we adopt an overlap scheduling structure based on the flooding decoding algorithm, which lets VNUs start iteration after a number of CNUs have been updated rather than all of them. The proposed PCM ensures that the elements required for subsequent column computations have been updated, as elaborated upon in Sect. 4.2.3. Through this approach, our decoder not only significantly enhances decoding efficiency and throughput, but also simplifies the scheduling logic and reduces logical resource consumption of the overlap controller.

4.2.3 Architecture of the Proposed QC-LDPC Code

The indices of the first-row elements within each submatrix of the original CCSDS (8176, 7154) PCM are randomized. When arranging all the elements within submatrices into a 511×511 matrix (as shown in Fig. 4.3a), the spacing between adjacent elements becomes randomized, while this random arrangement can yield greater coding gains, it presents challenges in applying the overlap scheduling method for further throughput enhancement during parallel decoding.

For instance, considering the original CCSDS (8176, 7154) PCM, the decoder might employ 16 VNU and 2 CNU for partial parallel iteration (where the degree of parallelism within submatrices is 1). This approach maximizes the utilization of its cyclic shift properties to save resources and increase throughput. Each VNU initiates computations in the rightwards direction from the position indexed as 1 in each column submatrix, until the current submatrix computation concludes. Therefore, while calculating VNU, it is essential that the column's corresponding elements have been updated with the latest information from CNU iteration.

In Fig. 4.3a, the red dots represent completed CNU iterations and updated information. If each submatrix's VNU starts iterating from the first column, it is necessary for all submatrices' first columns to have updated c2v messages before VNU iteration

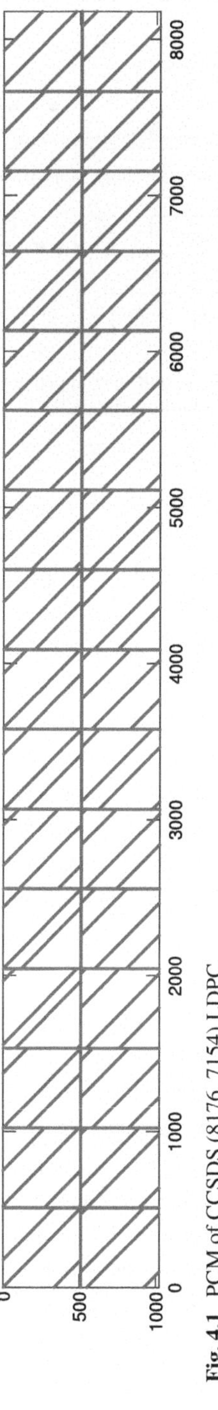

Fig. 4.1 PCM of CCSDS (8176, 7154) LDPC

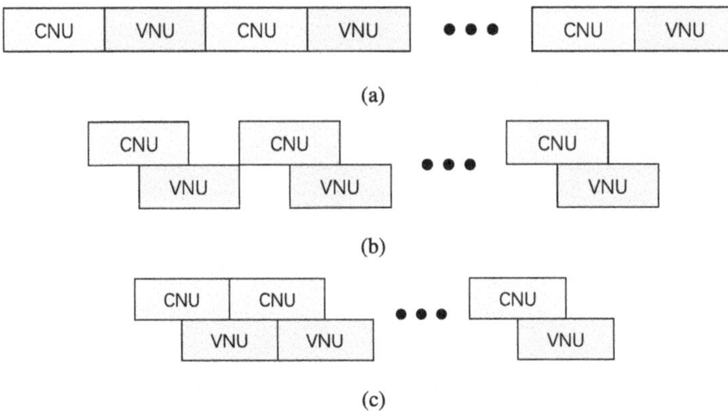

Fig. 4.2 Comparison of non-overlapped, half overlapped and full overlapped decoding scheme. **a** Non-overlapped; **b** Half overlapped; **c** Full overlapped

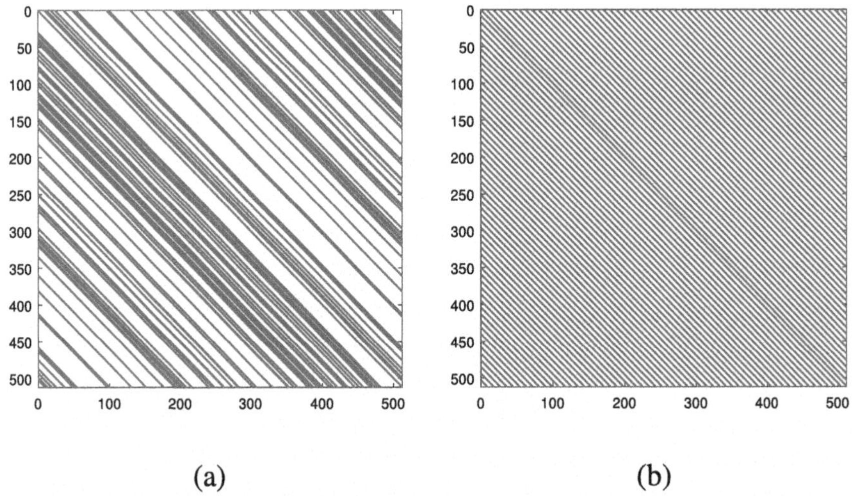

Fig. 4.3 Distribution of 1s in submatrices. **a** CCSDS (8176, 7154) LDPC code; **b** proposed (8176, 7154) LDPC code for overlap

commences. Through meticulous selection, if CNU begins iterating from row 459 (cycling back to 1 after reaching 511) and reaches row 416 (the completed iteration section in red), all submatrices' first columns will possess updated c2v messages, enabling the initiation of VNU iteration. Comparing this approach to the traditional flooding scheme, where VNU iteration must wait for CNU iteration completion, this method can let VNU start approximately 50 clock cycles earlier, before all CNU have been updated, which is also the overlap depth. Similarly, CNU can also be initiated after a certain number of VNUs have been iterated. Naturally, by selecting suitable starting points for VNU and CNU, or by using more intricate scheduling structures, the overlap depth can be extended, but introduces increased complexity in terms of hardware implementation as well.

Therefore, we propose an LDPC code based on CCSDS (8176, 7154) LDPC with a greater overlap depth, which is composed of 32 submatrices in 2 rows and 16 columns, and has submatrices of 511×511. In this LDPC code, the indices of first-row elements within each submatrix exhibit an approximately uniform distribution (the spacing between the last two index is smaller to fit in the 511×511 submatrix). When arranging all the elements within submatrices of this proposed PCM into a 511×511 matrix (as shown in Fig. 4.3b, the unfolded version of which can be referred to in Fig. 4.4), it becomes evident that the elements are nearly evenly spaced. This uniform distribution facilitates the design of more efficient and high-speed parallel overlap schemes. For the original CCSDS (8176, 7154) LDPC code, the characteristic where certain elements within specific matrices like $Q(s_{2,1})$, $Q(s_{2,3})$, ..., $Q(s_{2,15})$ have an index value of 1 is retained. For the remaining 56 elements, we introduce permutation vectors for the design process.

We employ a permutation vector shown in Algorithm (1) and defined in [26] to determine the shift factors (start index) of each submatrix, with parameters set to $m = 56$, $a = 30$, and $b = 40$, where m stands for the number of shift factors, and a, b are parameters defined after simulations to reach the best BER performance.

Algorithm 4.1 Permutation Vector.

Input: $[m, a, b]$
1: $l = m$
2: **for** $k1 = 1$ to m **do**
3:     $i = a * k1 + b \bmod(m + 1 - k1)$
4:     $\pi_A(k1) = m - s_i$
5:     **for** $k2 = i$ to $l - 1$ **do**
6:         $s_{k2} = s_{k2+1}$
7:     **end for**
8:     $l = l - 1$
9: **end for**
Output: π_A

Upon obtaining the permutation vectors, we proceed to uniformly arrange the data in the submatrices with a size of 511×511, according to the shift factors from the vectors. This is achieved by multiplying the permutation vectors by $\lfloor 511/56 \rfloor = 9$ to obtain the shift factors. Moreover, to enhance the BER performance, we employ a swapping scheme that compares adjacent factors originated by the permutation vector, and exchanges them with others if their distance is too small, thus maintaining a minimum interval of two diagonals. After these steps, the PCM we constructed is displayed in Fig. 4.4.

Utilizing the matrix we have proposed, it becomes remarkably convenient to design decoders with higher rates and lower resource consumption using the overlap scheme. When the submatrix parallelism level is set to 57, this approach can achieve convenient and regular deep overlap scheduling. The specific scheduling methodology will be discussed in subsequent sections.

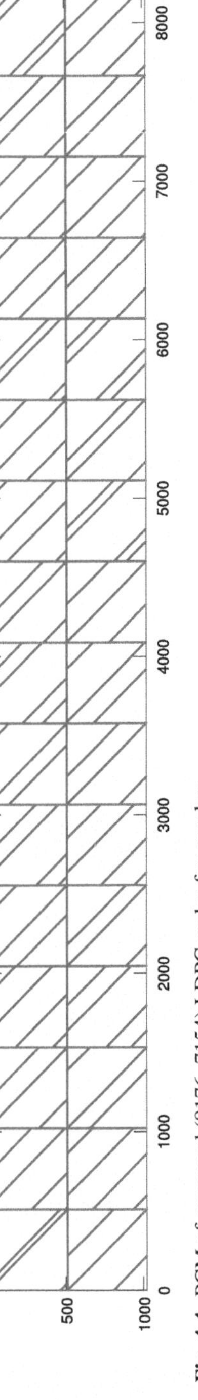

Fig. 4.4 PCM of proposed (8176, 7154) LDPC codes for overlap

4.3 Low-Complexity Decoding Algorithm

4.3.1 MSA

The process of LDPC decoding has been enhanced through the utilization of various algorithms, such as the Belief Propagation (BP) algorithm, Likelihood Ratios BP algorithm (LLR-BP), and the MSA [27–29]. The MSA, in particular, has garnered widespread adoption due to its ability to simplify hardware implementation without compromising decoding performance. Notably, Fossorier contributed to further streamlining the MSA's initialization steps by deriving formulas that enable the algorithm to decode without requiring prior knowledge of channel information [30]. In certain operations, direct assignment of initial probability values as inverses of received signal amplitudes enhances operational efficiency.

The complete MSA is shown as Algorithm (2), where rx, it_{max} stand for receiving message and the default iteration time, $L(q_n)$ stands for the soft message, $N(m)$ stands for nodes participating in the m-th check equation, $M(n)$ stands for check equations connected with the n-th variable node, $N(m)\backslash n$ stands for $N(m)$ without the n-th node, $M(n)\backslash m$ stands for $M(n)$ without the m-th equation, r_{mn} is the message from the m-th check equation to the n-th node, q_{mn} is the message form the i-th node to the j-th check equation, and $L(P[x_n \mid y, S])$ is the posteriori probability.

Algorithm 4.2 Min-Sum Algorithm.

Input: rx and PCM
1: **Initialization:**
2: $L(q_n) = -\text{rx}$
3: **Decoding:**
4: **while** t $\leq it_{max}$ and $Hx^T \neq 0$ **do**
5:    **CN Update:**
6:    **for** $i = 1 : m$ **do**
7:        $L(r_{mn}) = \prod_{n' \in N(m)\backslash n} \text{sign}(L(q_{mn'})) \cdot \min_{n' \in N(m)\backslash n} abs(L(q_{mn'}))$
8:    **end for**
9:    **VN Update:**
10:    **for** $j = 1 : n$ **do**
11:        $L(q_{mn}) = L(q_n) + \sum_{m' \in M(n)\backslash m} L(r_{m'n})$
12:        **Posterior probability Update:**
13:        $L(P[x_n \mid y, S]) = L(q_j) + \sum_{m \in M(n)} L(r_{mn})$
14:        **Decoding decision:**
15:        **if** $L(P[x_n \mid y, S]) > 0$ **then**
16:            $x_n = 0$
17:        **else**
18:            $x_n = 1$
19:        **end if**
20:    **end for**
21:    $t = t + 1$
22: **end while**
Output: Decoded Data x

4.3.2 Modified 2-Bit MSA

In order to implement high-speed hardware for LDPC decoding, it is necessary to quantize the message values. Quantization with fewer bits can save memory and simplify the hardware structure, but it may also lead to a reduction in BER performance [31]. Based on the original MSA, [22] gave a modification only using 2-bit quantization, while retaining a good decoding performance, but greatly simplifies the computation and storage. In this algorithm, intrinsic message is quantized using 2 bits, and is represented by $B_s B_m$ [32]. Unlike traditional quantization methods, the 2-bit MSA does not directly represent the true values of the messages, but instead represents the confidence coefficient. B_s represents a hard decision of the received message, while B_m indicates the reliability of the hard decision [33]. When $B_m = 0$, it represents a low confidence level but not the value of zeros, while $B_m = 1$ indicates a high level of confidence in the hard decision. The values of B_s and B_m can be obtained using Eqs. (4.1) and (4.2), where rx is the received message and T_y is a transformation threshold. The value of T_y is determined as 4/10 through simulations.

$$B_s = \text{sign}(rx), \tag{4.1}$$

$$B_m = \begin{cases} 1, & \text{if } |rx| > T_y, \\ 0, & \text{otherwise} . \end{cases} \tag{4.2}$$

After receiving the message rx, the initialization part first transforms rx from integers into 2-bit messages according to Eq. (4.3), where I_n stands the intrinsic message converted from rx.

$$I_n = B_s B_m = \begin{cases} 01, & \text{if } rx > T_y, \\ 00, & \text{if } T_y \geq rx \geq 0, \\ 10, & \text{if } 0 > rx \geq -T_y, \\ 11, & \text{if } rx < -T_y. \end{cases} \tag{4.3}$$

When implementing on FPGA, direct use of adders or comparators consumes more LUT resources compared to combinational logic circuits. To enhance the HUE and to adapt to the dual-port characteristics of the gate circuit, we have devised an improved algorithm that builds on these observations.

To optimize the memory structure, in CNU, only 3 bits are used to store each c2v message, representing the product of sign bits, min_1 and min_2, instead of storing all the R_{mn} results. The values of min_1 and min_2, which are calculated in CNU, can be used to restore the exact R_{mn} in VNU before use [22]. Due to the fact that the magnitude part of the CNU's input is only 1 bit, finding min_1 and min_2 can be easily formulated as Boolean operations with AND and OR gates. For example, the expression for finding min_1 and min_2 for 8 bits is given by Eq. (4.4).

$$
\begin{aligned}
a &= B_{m1}B_{m2}B_{m3} + B_{m1}B_{m2}B_{m4} + B_{m1}B_{m3}B_{m4} + B_{m2}B_{m3}B_{m4} \\
&= B_{m1}B_{m2}(B_{m3} + B_{m4}) + (B_{m1} + B_{m2})B_{m3}B_{m4}, \\
b &= B_{m5}B_{m6}B_{m7} + B_{m5}B_{m6}B_{m8} + B_{m5}B_{m7}B_{m8} + B_{m6}B_{m7}B_{m8} \\
&= B_{m5}B_{m6}(B_{m7} + B_{m8}) + (B_{m5} + B_{m6})B_{m7}B_{m8}, \\
c &= B_{m1}B_{m2}B_{m3}B_{m4}, \\
d &= B_{m5}B_{m6}B_{m7}B_{m8}, \\
min_1 &= cd, \\
min_2 &= abc + abd + acd + bcd = ab(c + d) + (a + b)cd.
\end{aligned}
\tag{4.4}
$$

To optimize the hardware implementation of the CNUs, a pipeline operation can be performed to divide the 32 inputs into four groups. This enables the calculation of four sets of min_1 and min_2, which can then be cascaded to a second stage by inputting these sets. By doing so, the min_1 and min_2 of all 32 B_m can be obtained efficiently. This approach effectively reduces the hardware resources required for the CNU implementation, while ensuring high performance and accuracy of the decoder.

There are four steps in VNU. First, four 3-bit c2v messages are inputted based on the column weight. Next, four 2-bit R_{mn} are recovered with the help of $L_n^{(k-1)}$, which means the L_n in the $(k-1)$-th iteration. Then, these four R_{mn} and the I_n will be converted into integers according to Eq. (4.6), which will be added together later. At last, the summation will be converted back to 2 bits and saved as v2c. Before going into more detail about VNU, two auxiliary functions will be introduced first. Function $f(\cdot)$ and $g(\cdot)$ are used in the VNU to convert 2-bit messages into integers for addition operation, and to transform the integer results back into 2-bit messages for storage, respectively. Specifically, $f(\cdot)$ is formulated in Eq. (4.5) and $g(\cdot)$ is formulated in Eq. (4.6). The magnitude bit of the integer result in $f(\cdot)$ corresponds to the confidence coefficient B_m, with a value of w for $B_m = 0$ indicating low confidence, and a value of W for $B_m = 1$ indicating high confidence. These values are determined through simulations, along with the threshold T_y for $f(\cdot)$ and T_L for the VNU process.

$$
f(B_s B_m) = \begin{cases}
w, & \text{if } B_s B_m = 00, \\
W, & \text{if } B_s B_m = 01, \\
-w, & \text{if } B_s B_m = 10, \\
-W, & \text{if } B_s B_m = 11.
\end{cases}
\tag{4.5}
$$

$$
g(x) = \begin{cases}
00, & \text{if } x \geq T_L, \\
01, & \text{if } T_L > x > 0, \\
10, & \text{if } 0 > x > -T_L, \\
11, & \text{if } x \leq -T_L.
\end{cases}
\tag{4.6}
$$

And the origin computation in VNU can formulated as Eq. (4.7) with $f(\cdot)$ and $g(\cdot)$.

$$L_{mn} = L_n = g\left(f(I_n) + \sum_{j \in M(n)} f(R_{jn}) \right). \tag{4.7}$$

Literatrue [34] presents a method to recover the exact R_{mn} using $L_n^{(k-1)}$, $\min_1$, and $\min_2$, which is described in Eq. (4.8), where $S_m^{(k)}$ represents the sign product of nodes connected with the m-th check equation in the k-th iteration.

$$\left| R_{mn}^{(k)} \right| = \min 1_m^{(k)} + \min 2_m^{(k)} \overline{\left| L_n^{(k-1)} \right|},$$
$$\mathrm{sign}\left(R_{mn}^{(k)} \right) = S_m^{(k)} \oplus \mathrm{sign}\left(L_n^{(k-1)} \right). \tag{4.8}$$

Our w and W in $f(\cdot)$ are chosen as 1 and 5, so we need four bits to represent the four potential integers, which are $-5, -1, 1, 5$. Furthermore, during the conversion, we summarized the following connection between c2v messages' two bits (B_s, B_m) and integers' four bits (b_1, b_2, b_3, b_4) as Eq. (4.9) shows. In this way, we are able to substitute simple logic circuit for comparers.

$$\begin{aligned}
b_1 &= B_s, \\
b_2 &= B_s \oplus B_m, \\
b_3 &= B_s, \\
b_4 &= 1.
\end{aligned} \tag{4.9}$$

The last step is the summation, which we realize using a 2-stage pipeline. At last, the first bit of the result, which is the sign bit, will be kept as B_s, and the absolute value of the result will be compared with T_L and converted into B_m. B_s and B_m will make up the $L_n^{(k)}$, which is also the v2c messages.

4.4 Computation Unit Design for the Decoder

4.4.1 System Architecture

The architecture of our FPGA-based LDPC decoder is presented in Fig. 4.5. It consists of several components, including CNU, VNU, shift-register-based memory, overlap controller, and hard decision module. The system utilizes a partial parallel architecture with a degree of parallelism of $2 \times 57 = 114$ for the CNU and $16 \times 57 = 912$ for the VNU.

The received soft message is quantified and truncated to 2 bits, and then stored in the shift-register based memory, which will be discussed in detail in Sect. 4.5.1. With the help of Modified 2-bit MSA, the memory efficiency is improved, and up to two thirds of memory resources can be saved.

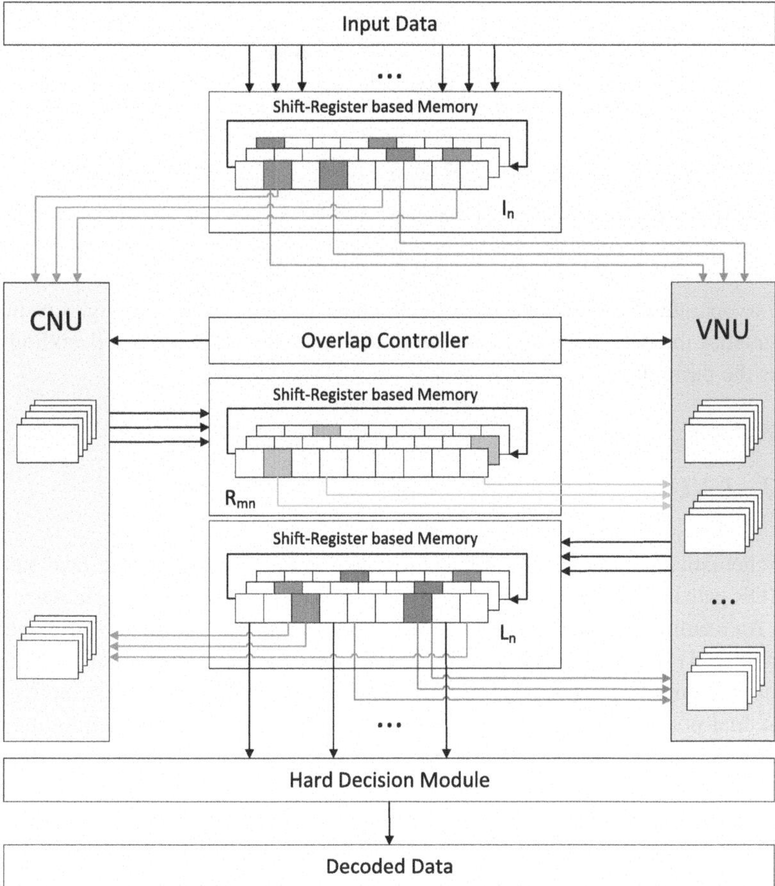

Fig. 4.5 Schematic diagram of decoding system

After CNU operation, a c2v information consisting of 3 bits is generated. These 3 bits are the product of all symbol bits calculated by the XOR tree, as well as the minimum and the subminimum value of all amplitude bits. It is worth noting that in our decoder, the computation of the minimum and the subminimum values is achieved through logical operations, rather than comparators, which leads to a smaller computational delay. Furthermore, since all gate circuits are dual-ported, we further simplify the logical expression, thereby simplifying the circuit and reducing resource overhead. A detailed explanation of CNU will be provided in Sect. 4.4.2.

Once enough rows have been iterated, the overlap controller sends the synchronization signal to VNU, and the v2c result is derived after the conversion and addition. The specific operation in VNU will be introduced in Sect. 4.4.3. Both c2v and v2c messages are stored in the shift-register-based memory as well.

Hard decisions begin when the number of iterations reaches the preset value. Using the proposed QC-LDPC matrix optimized for overlap, the overlap controller allows

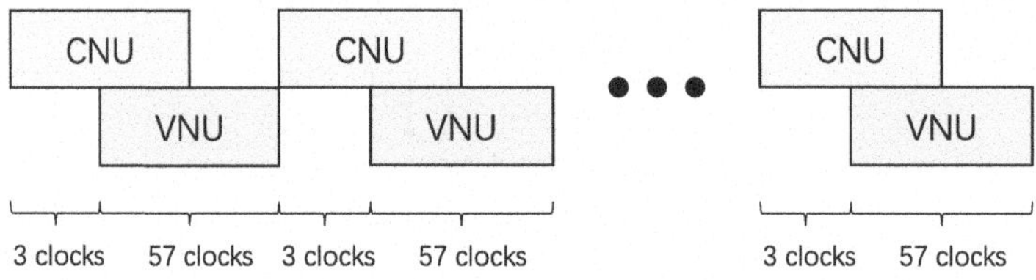

Fig. 4.6 Overlap timing relationship in decoding

VNU to operate when three groups of CNU have been updated, like Fig. 4.6 shows, and satisfies the data dependency at the same time, which accelerates decoding and triples the throughput.

4.4.2 CNU

The schematic diagram of the CNU is presented in Fig. 4.7. In the gate-level netlist, the XOR gate is dual-ported, which allows for a five-stage XOR tree as shown in Fig. 4.7a. An example of the module used to determine the min_1 and min_2 of eight bits based on (4.4) is provided in Fig. 4.7b. To obtain the min_1 and min_2 of all 32 bits, the output of this module is cascaded and the eight resulting values are used as inputs. The use of dual-ported characteristics simplifies the netlist, resulting in a requirement of only 239 nets and 175 cells in Vivado implementation, which is a significant reduction in resources, amounting to two-fifths of the original requirement.

The VNU's schematic is depicted in Fig. 4.8, where the numbers after the slashes represent their corresponding bit widths. Firstly, four R_{mn} values are generated by utilizing c2v and $L_n^{(k-1)}$, which will later be converted to integers. The crucial step in VNU involves adding five integers within a two-stage pipeline adder, yielding a 7-bit result that will subsequently be transformed back into 2 bits as Eq. (4.6). The first bit is stored directly as B_s, while the remaining 6 bits are taken as absolute values based on the first bit. The absolute value is then compared to T_L, obtaining the B_m that will be saved as v2c, along with B_s. Through a Modified 2-bit MSA and substituting partial comparators with logic circuits, the netlist is simplified, necessitating a mere 68 cells and 122 nets for each VNU in Vivado implementation.

4.4.3 VNU

The schematic of VNU is presented in Fig. 4.8 where the values after slashes are the width of the number. Four R_{mn} are generated first with c2v and $L_n^{(k-1)}$, which will be converted into integer later. And then is the key step in VNU, five integers are added within a 2 stage pipeline adder, and the result's width is 7 bits, which will be converted back to 2 bits according to (4.6). The first bit is saved directly as B_s, and

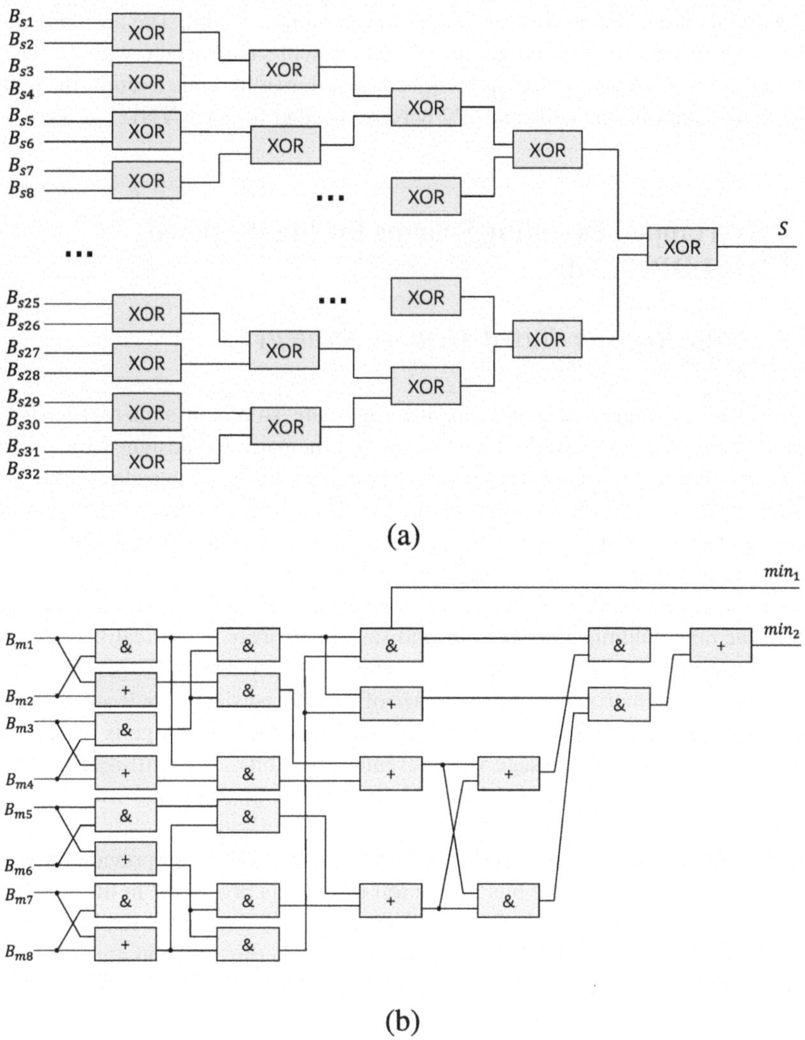

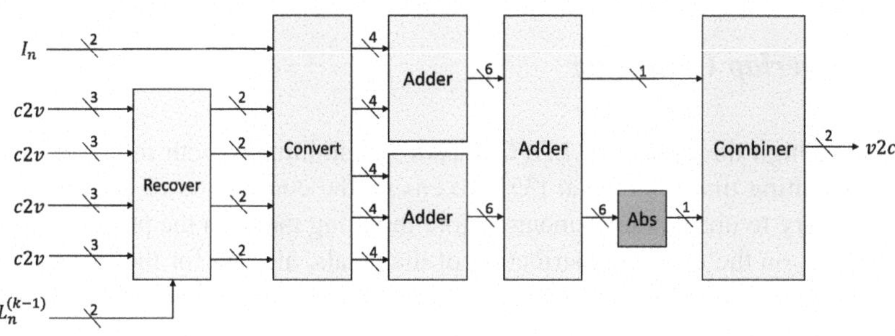

Fig. 4.7 Schematic of check node unit. **a** XOR tree of B_s; **b** operation of B_m

Fig. 4.8 Schematic of variable node unit

the rest 6 bits are taken as absolute values based on the first bit. The absolute value will be compared with T_L, and get the B_m which will saved as v2c with B_s. Using Modified 2-bit MSA and replacing partial comparators with logic circuits, the netlist is facilitated and only 68 cells and 122 nets are needed for each VNU.

4.5 Overlapped Decoding Scheme for the Proposed QC-LDPC Code

4.5.1 Shift-Register-Based Memory Strategy

All intrinsic messages, c2v, and v2c messages are stored in shift-register-based memory units instead of RAMs. The memory schematic we designed for overlap is depicted in Fig. 4.9. Each row produces one c2v message per clock, and each column produces one v2c message per clock. Therefore, the data are divided into vectors and saved as an array according to the size of the submatrix in the order of rows (columns) without address conflicts.

Since the degree of parallelism is 57, we divide elements in one submatrix, dividing each of the nine columns into a group, and save in an array. Specifically, the size of registers storing the intrinsic message is $[2(\text{width}) \times 9(\text{length})] \times 57(\text{parallelism}) \times 16(\text{number of submatrix in a row})$. The size of the v2c register is $3 \times 9 \times 57 \times 2$, the size of the c2v register is $2 \times 9 \times 57 \times 16$, since each c2v message is represented by 3 bits, and each v2c message is represented by 2 bits. The shift-register-based memory unit cyclic shifts 2 or 3 bits and updates the first few bits with the latest calculation results.

In QC submatrices, every row (or column) is shifted by the previous row (or column). This means that the bits on the left of the bits processed in this clock are the bits needed for the next clock. In this way, once we decide the read port of the vector, with the cyclic shift of the register, the bits of this position are always the pending ones. This reduces the routing complexity.

Furthermore, the storage method with a fixed read/write position can eliminate the read/write latency and further accelerate the decoding process.

4.5.2 Overlap Controller

To achieve high throughput in LDPC decoding, minimizing both intra- and inter-iteration waiting times is crucial [35]. To ensure the correctness of each iteration, it is necessary to update the elements before inputting them. In the proposed PCM, which relies on the uniform distribution of diagonals, all data for the first variable node will have finished updating after the first three row processes. This allows the VNU to continue processing because the elements are saved in shift-registers.

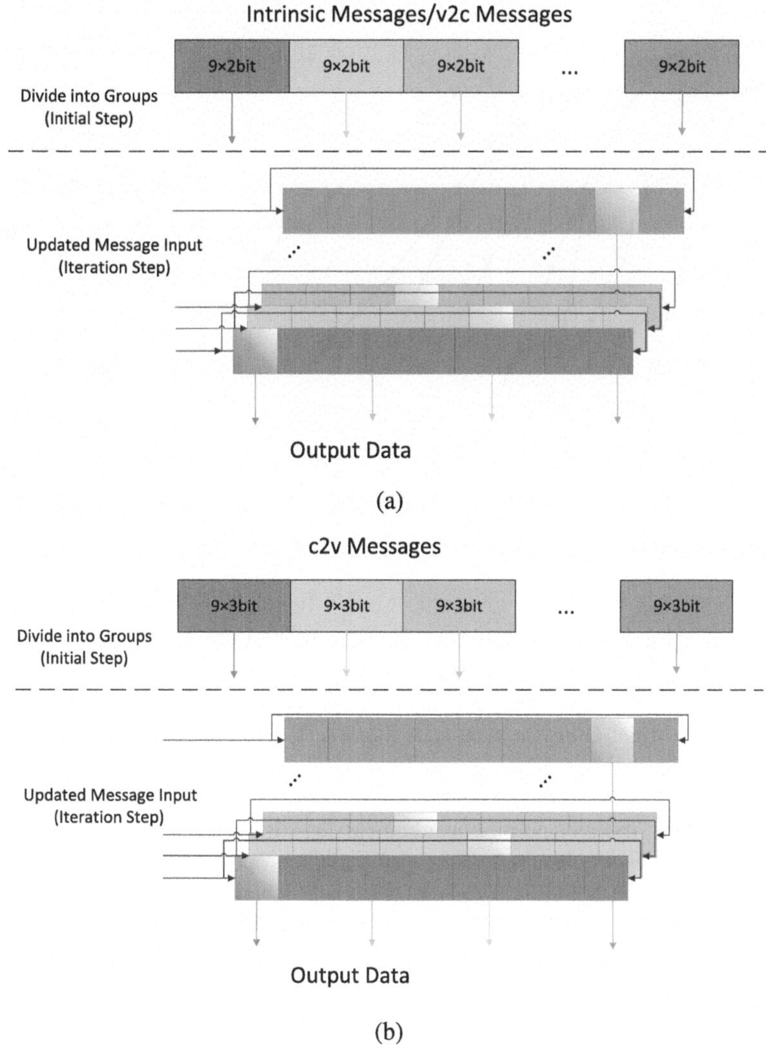

Fig. 4.9 Schematic of check node unit. **a** Shift-register-based memory for intrinsic message and v2c message; **b** shift-register-based memory for c2v message

4.6 Results and Discussion

4.6.1 Simulation Results

The method to determine the parameters is finding the value with best BER performance at the same SNR. After simulations in MATLAB, we determined the values of T_y, w, W, and T_L to be 4/10, 1, 5, and 4, respectively. The BER performance

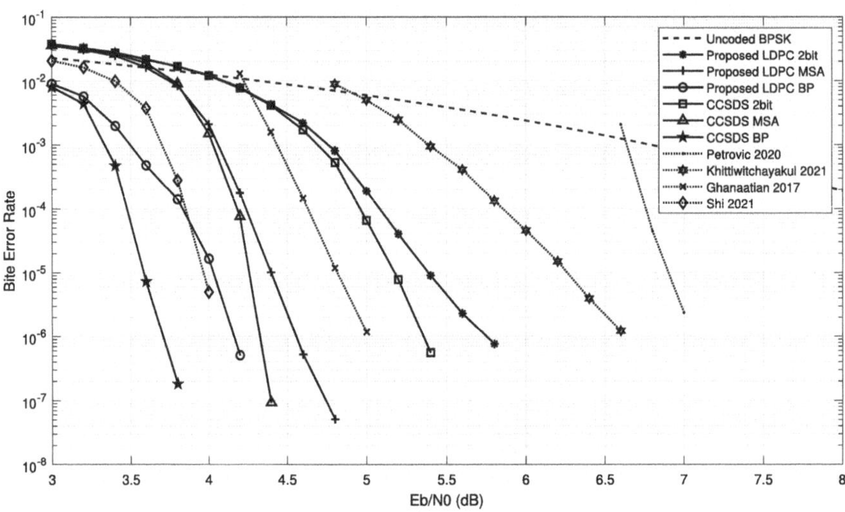

Fig. 4.10 Bit error rate analysis and performance of decoders in [36–39]

simulation of proposed low-complexity high-throughput QC-LDPC decoder architecture for proposed LDPC was carried out using BPSK modulation in an AWGN channel. Figure 4.10 illustrates the BER performance of our implementation.

The six solid lines show the results of the MATLAB simulations. Three distinct decoding algorithms were employed, each executed over eight iterations, for both the CCSDS (8176, 7154) LDPC and the proposed LDPC code, yielding comprehensive simulation results. Initially, employing identical decoding algorithms, a minor coding gain loss of 0.5 dB was observed when applying the proposed LDPC in comparison to the original CCSDS (8176, 7154) LDPC code. This outcome underscores the effectiveness of our devised PCM, which yields a favorable performance comparable to the original CCSDS (8176, 7154) LDPC, and this 0.5 dB coding gain loss is acceptable considering that the throughput can be significantly improved by using the proposed LDPC with the same resource consumption. And, it can be found that no matter which PCM is used, there is a 1 dB coding gain loss using 2-bit MSA compared to MSA when the BER reaches 10^{-6}. This trade-off, albeit resulting in a BER performance compromise, remains acceptable, as the 2-bit MSA significantly reduces hardware resources and complexity, all while achieving approximately two-thirds reduction in resource overhead. Overall, our decoder simultaneously increases throughput and reduces decoding overhead by sacrificing BER performance.

Meanwhile, we have selected four decoders with high throughput or low implementation complexity as the design principle and similar code rate, and compared the BER performance; the results are shown by four dashed lines in Fig. 4.10. The decoder in [36] uses the (10368, 8448) LDPC from 5G NR with a code rate of 22/27, performs 10 iterations, employs 8-bit quantization, and the algorithm used is a layered decoding algorithm. It achieves a BER of 10^{-6} at a Eb/N0 greater than

7. In comparison, this decoder has higher throughput and HUE, but obviously the coding gain for the same SNR is not as good as our decoder. The decoder in [37] uses the (2048, 1723) LDPC from IEEE 802.3an with a code rate of about 7/8, five iterations, 4-bit quantization, and the decoding algorithm used is MSA. It has a SNR of 5.6 at a BER of 10^{-6}. This decoder has extremely high throughput, and its BER performance is slightly better than ours, but its fully parallel architecture leads to significant resource overhead. The decoder in [38] uses the constructed (4608, 4096) LDPC with a code rate of 8/9, performs six iterations, uses the decoding algorithm RR-BP proposed in that article, and the encoding and decoding process are both unquantized. Its BER achieves 10^{-6} at a Eb/N0 of 6.7. There is no information about the implementation of this decoder, but it is clear that its BER performance lags behind our decoder by almost 2 dB. The decoder of [39] uses CCSDS (8176, 7154) LDPC with code rate 7/8, performing 20 iterations, the decoding algorithm used is NMS, and the encoding and decoding process are both unquantized. Its BER achieves 10^{-6} at an Eb/N0 of 3.9. The NMS used by this decoder has about 0.5 dB performance improvement compared to MSA, but due to our deep overlap structure, our decoder has much higher HUE and we have higher throughput with the same resource consumption. We designed the overlap decoder using 2bit MSA, with low resource consumption and high throughput as the design criterion. Compared to other decoders, which either have high rate but high resource consumption or low resource consumption but also low rate, our decoder can achieve a balance between resource and rate, which can meet the demand of low power consumption and high throughput well, and at the same time has a good BER performance.

As an additional facet of our analysis, Fig. 4.11 provides insights into the average number of iterations required for successful decoding, which shows the convergence speed of the algorithms. This figure reaffirms the slight disparity introduced by our proposed LDPC code when compared to the CCSDS LDPC code at higher SNRs. Furthermore, it is evident that utilizing the 2-bit MSA leads to a performance penalty, underscoring the trade-off inherent in its application.

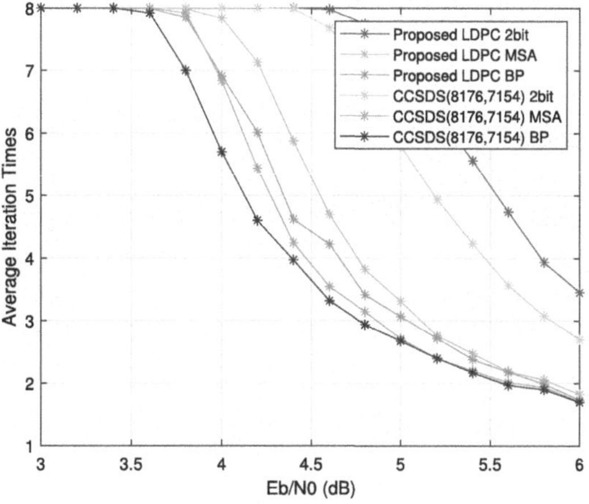

Fig. 4.11 Number of iterations analysis

4.6.2   Experimental Results

For our design, we used Xilinx Virtex UltraScale+ FPGA VCU118 Evaluation Board (XCVU 9P) as the implementation platform. The throughput of a decoder is often a crucial performance metric, and it can be calculated using the following formula:

$$T_{\max} = \frac{r \times l \times f}{n \times c}, \tag{4.10}$$

where r is the code's rate, l is the code's length, f is the frequency of the decoder, n is the number of iterations, and c is the number of clocks per iteration [40].

For our optimized decoder designed for overlap, we can finish a single iteration in 18 clocks. Suppose we use a clock frequency of 156.25 MHz and perform eight iterations, which is a common choice for high-performance LDPC decoders. The code we use has a rate of 0.875, a length of 8176 and thus, the throughput of the decoder can be calculated as follows:

$$T_{\max} = \frac{0.875 \times 8176 \times 156.25 \text{ MHz}}{8 \times 18} = 7.76 \text{ Gbps}. \tag{4.11}$$

Therefore, our optimized decoder can achieve a throughput of 7.76 Gbps, which is quite high and suitable for high-speed systems. If overlap control is not used under the same decoder, one iteration takes 114 clocks, resulting in a throughput of 4.08 Gbps. In other words, the overlap decoding scheme doubles throughput while maintaining unchanged resources and performance.

The utilization of the FPGA is presented in Table 4.1. From the table, we can find that we only need 31 LUTs per CNU and consume only 20 LUTs per VNU, which is minimal for accomplishing the comparison of 32 elements as well as the transformation and summation processing of five elements. The key to this is that we implement functions such as comparison, addition, and mapping of data through combinatorial logic, and simplify the logical relationships as much as possible. It is important to note that RAMs are not considered in this calculation, as they are not used in the shift-register-based memory scheme employed in our implementation.

Table 4.1 Hardware utilization

Computation unit	LUT	Register	RAM	DSP
CNU	$31 \times 2 \times 57 = 3534$	$26 \times 2 \times 57 = 2964$	0	0
VNU	$20 \times 16 \times 57 = 18,240$	$28 \times 16 \times 57 = 25,536$	0	0
Permutation net	51,944	114,004	0	0
Total	73,718	142,504	0	0
Available	1,182,240	236,4480	2160	6840

The permutation net, including storage, requires a total of 73,718 LUTs and 142,504 registers. Overall, our decoder utilizes only 10% of the total resources on our board. This low resource usage is a testament to the efficiency of our implementation, and demonstrates its practical feasibility for real-world applications.

Furthermore, a comparison of the implementation and performance of the proposed low-complexity high-throughput decoder against various other LDPC decoders with high HUE or similar PCM is presented in Table 4.2.

The findings detailed in Table 4.2 underscore the commendable performance of the proposed decoder in comparison to other reported decoding techniques. This achievement is particularly notable in terms of achieving a high throughput while operating at a lower working frequency. Noteworthy metrics such as Mbps/kLUT, Mbps/kFF, and Mbps/36kb BRAM are employed for comprehensive comparison, highlighting the different HUE aspects. Our Mbps/kLUT ratio emerges as highly competitive, showcasing the efficiency of our design. However, the Mbps/kFF ratio does not distinctly favor our approach, mainly due to our utilization of shift-register-based memory as opposed to BRAMs. The high throughput is mainly achieved by relying on deep overlap with the proposed LDPC. Simultaneously, the achievement of high HUE is supported by the adoption of the 2-bit MSA and the utilization of combinational logic-based CNUs and VNUs. Evidently, the implementation results elucidate the efficacy of the design methodology proposed in this study. Its attributes of low complexity, high HUE, and impressive throughput render it a fitting solution, particularly apt for applications in sensor networks.

Table 4.2 Implementation and comparison of different decoders

	Proposed decoder	[36]	[16]	[41]	[42]
Standard	Proposed LDPC	5G NR	5G NR	CCSDS	CCSDS
Code rate	7/8	22/27	1/3	7/8	7/8
Code length	8176	10368	6528	8176	8176
Max. Iter.	8	5	10	10	10
Algorithm	Modified 2-bitMSA	HybridSchedule	OMS	F-NMS	F-AMSA
LLRs Quant.	2	8	5	7	6
Throughput (Gbps)	7.76	31.7	2.168	2	1.02
Frequency (MHz)	156.25	261	82	250	250
LUT	73,718	100,929	225,191	56,778	46,294
FFs	142,504	85,431	–	86,942	39,103
BRAM (KB)	0	4896	3456	573	253
Mbps/kLUT	106.3	314.2	9.6	35.7	22.2
Mbps/kFF	54.6	371.3	–	23	26.2
Mbps/36kb BRAM	–	232.4	22.6	125	145

4.7 Conclusions

This chapter presents a novel overlap-optimized decoder for high-speed communications, aimed at fulfilling the high data rate requirements of this domain. A QC-LDPC code is constructed, which allows for decoding time overlap of up to one-third, thereby enhancing the decoder's throughput. To strike a balance between decoding performance and implementation complexity, a Modified 2-bit MSA is used. In addition, a Boolean operation-based CNU and VNU are proposed to facilitate c2v/v2c message calculation by logic circuits, thereby reducing resource overhead. To address the quasi-cyclic characteristic and reduce read/write latency, a shift-register-based memory strategy is employed. The proposed decoder achieves a coding gain of 5 dB at a BER of 10^{-6} and attains a throughput of 7.76 Gbps at a working frequency of 156.25 MHz. Implementation results on the Xilinx Virtex UltraScale + FPGA VCU118 Evaluation Board reveal that the proposed decoder uses only 10% of the total resources of the board, attesting to its efficiency and practicality. In summary, the proposed overlap-optimized decoder provides a promising solution to meet the high data rate requirements of sensor networks.

References

1. R. Gallager, Low-density parity-check codes. IRE Trans. Inform. Theory **8**(1), 21–28 (1962)
2. L. Zhang, Development and prospect of chinese lunar relay communication satellite, *Space: Science & Technology*, no. 001 (2021)
3. Q. Meng, M. Huang, Y. Xu, N. Liu, X. Xiang, Decentralized distributed deep learning with low-bandwidth consumption for smart constellations, *Space: Science & Technology*, no. 001, (2021)
4. S.-Y. Chung, G.D. Forney, T.J. Richardson, R. Urbanke, On the design of low-density parity-check codes within 0.0045 db of the shannon limit. IEEE Commun. Lett. **5**(2), 58–60 (2001)
5. M. Shi, K. Yang, D. Niyato, H. Yuan, H. Zhou, Z. Xu, The meta distribution of sinr in uav-assisted cellular networks. IEEE Trans. Commun. **71**(2), 1193–1206 (2023)
6. J.R. Barry, Low-density parity-check codes. Georgia Inst. Technol. **5**, 1–20 (2001)
7. X. Chen, J. Kang, S. Lin, V. Akella, Memory system optimization for fpga-based implementation of quasi-cyclic ldpc codes decoders. IEEE Trans. Circuits Syst. I Regular Papers **58**(1), 98–111 (2010)
8. V.A. Chandrasetty, S.M. Aziz, Resource efficient ldpc decoders for multimedia communication. Integration **48**, 213–220 (2015)
9. N. Andreadou, F.-N. Pavlidou, S. Papaharalabos, P.T. Mathiopoulos, Quasi-cyclic low-density parity-check (qc-ldpc) codes for deep space and high data rate applications, in *2009 International Workshop on Satellite and Space Communications*, pp. 225–229 (2009)
10. Y. Wang, J. Yedidia, S. Draper, Construction of high-girth qc-ldpc codes, in *2008 5th International Symposium on Turbo Codes and Related Topics*, pp. 180–185 (2008)
11. CCSDS-131.1-O-2, Low density parity check codes for use in near-earth and deep space applocations[r], Washington, DC, USA (2007)
12. J. Lou, Y. Jiao, R. Yang, Y. Huang, X. Xu, L. Zhang, Z. Ma, Y. Yu, W. Peng, Y. Yuan et al., Calibration-free, high-precision, and robust terahertz ultrafast metasurfaces for monitoring gastric cancers. Proc. Nat. Acad. Sci. **119**(43), e2209218119 (2022)

13. J. Lou, X. Xu, Y. Huang, Y. Yu, J. Wang, G. Fang, J. Liang, C. Fan, C. Chang, Optically controlled ultrafast terahertz metadevices with ultralow pump threshold. Small **17**(44), 2104275 (2021)
14. X. Xu, J. Lou, S. Wu, Y. Yu, J. Liang, Y. Huang, G. Fang, C. Chang, Snse 2-functionalized ultrafast terahertz switch with ultralow pump threshold. J. Mater. Chem. C **10**(15), 5805–5812 (2022)
15. N. Ye, X. Cao, X. Ding, J. Li, D. Zhao, Q. Ouyang, Multi-connection to the sky: energy-efficient beamforming for multi-satellite uplink transmission with lens antenna array, *IEEE Transactions on Green Communications and Networking*, pp. 1–1 (2023)
16. S. Pourjabar, G.S. Choi, A high-throughput multi-mode ldpc decoder for 5g nr. arXiv preprint arXiv:2102.13228 (2021)
17. Q. Lu, C.W. Sham, F. Lau, Rapid prototyping of multi-mode qc-ldpc decoder for 802.11n/ac standard, in *Asia & South Pacific Design Automation Conference* (2016)
18. S.-S. Yang, J.-Q. Liu, Z.-G. Lu, Z.-L. Bai, X.-Y. Wang, Y.-M. Li, An fpga-based ldpc decoder with ultra-long codes for continuous-variable quantum key distribution. IEEE Access **9**, 47687–47697 (2021)
19. E. Yeo, P. Pakzad, B. Nikolic, V. Anantharam, Vlsi architectures for iterative decoders in magnetic recording channels. IEEE Trans. Magnet. **37**(2), 748–755 (2001)
20. Y. Chen, K.K. Parhi, Overlapped message passing for quasi-cyclic low-density parity check codes. IEEE Trans. Circuits Syst. I Regular Papers **51**(6), 1106–1113 (2004)
21. Z. Wang, Z. Cui, Low-complexity high-speed decoder design for quasi-cyclic ldpc codes. IEEE Trans. Very Large Scale Integr. Syst. **15**(1), 104–114 (2007)
22. Z. Cui, Z. Wang, Improved low-complexity low-density parity-check decoding. IET Commun. **2**(8), 1061–1068 (2008)
23. Fossorier, M. Pc, Quasicyclic low-density parity-check codes from circulant permutation matrices. IEEE Trans. Inform. Theory **50**(8), 1788–1793 (2004)
24. T. Nguyen, T.N. Tan, H. Lee, Low-complexity high-throughput qc-ldpc decoder for 5g new radio wireless communication. Electronics **10**(4), 516 (2021)
25. S.M. Kim, K.K. Parhi, Overlapped decoding for a class of quasi-cyclic ldpc codes, in *IEEE Workshop on Signal Processing Systems* (2004)
26. R. Echard, S.C. Chang, Design considerations leading to the development of good π-rotation ldpc codes **9**(5), 449 (2005)
27. D.J. MacKay, R.M. Neal, Good codes based on very sparse matrices, in *IMA International Conference on Cryptography and Coding*, pp. 100–111, Springer (1995)
28. D. Envelope, R.S. Selvakumari, Performance analysis of min-sum based ldpc decoder architecture for 5g new radio standards (2022)
29. H. Hatami, D. Mitchell, D.J. Costello, T.E. Fuja, A threshold-based min-sum algorithm to lower the error floors of quantized ldpc decoders. IEEE Trans. Commun. **PP**(99), 1 (2020)
30. M. Fossorier, M. Mihaljevic, H. Imai, Reduced complexity iterative decoding of low-density parity check codes based on belief propagation. IEEE Trans. Commun. **47**(5), 673–680 (2002)
31. R. Zarubica, R. Hinton, S.G. Wilson, E.K. Hall, Efficient quantization schemes for ldpc decoders, in *Military Communications Conference, 2008. MILCOM 2008. IEEE* (2008)
32. V.A. Chandrasetty, S.M. Aziz, Fpga implementation of high performance ldpc decoder using modified 2-bit min-sum algorithm, in *2010 Second International Conference on Computer Research and Development*, pp. 881–885, IEEE (2010)
33. V.A. Chandrasetty, S.M. Aziz, An area efficient ldpc decoder using a reduced complexity min-sum algorithm. Integration **45**(2), 141–148 (2012)
34. Camille, Leroux, Saied, Hemati, Shie, Mannor, Warren, J., Gross, Kevin, High-throughput energy-efficient ldpc decoders using differential binary message passing, *IEEE Transactions on Signal Processing: A publication of the IEEE Signal Processing Society*, vol. 62, no. 3, pp. 619–631 (2014)
35. Y. Dai, Z. Yan, N. Chen, Optimal overlapped message passing decoding of quasi-cyclic ldpc codes, *IEEE Transactions on Very Large Scale Integration (VLSI) Systems* (2008)

36. V.L. Petrović, M.M. Marković, D.M. El Mezeni, L.V. Saranovac, A. Radošević, Flexible high throughput qc-ldpc decoder with perfect pipeline conflicts resolution and efficient hardware utilization. IEEE Trans. Circuits Syst. I Regular Papers **67**(12), 5454–5467 (2020)
37. R. Ghanaatian, A. Balatsoukas-Stimming, T.C. Müller, M. Meidlinger, G. Matz, A. Teman, A. Burg, A 588-gb/s ldpc decoder based on finite-alphabet message passing. IEEE Trans. Very Large Scale Integr. (VLSI) Syst. **26**(2), 329–340 (2017)
38. S. Khittiwitchayakul, W. Phakphisut, P. Supnithi, Reliability ratio-based serial algorithm of ldpc decoder for turbo equalization schemes. IEEE Trans. Magnet. **58**(2), 1–5 (2021)
39. H. Shi, M. Zhong, Z. Luo, C. Li, Low complexity neural network-aided nms ldpc decoder, in *2021 Computing, Communications and IoT Applications (ComComAp)*, pp. 227–231, IEEE (2021)
40. K. Zhang, X. Huang, Z. Wang, High-throughput layered decoder implementation for quasi-cyclic ldpc codes. IEEE J. Sel. Areas Commun. **27**(6), 985–994 (2009)
41. T. Xie, B. Li, M. Yang, Z. Yan, Memory compact high-speed qc-ldpc decoder, in *2017 IEEE International Conference on Signal Processing, Communications and Computing (ICSPCC)*, pp. 1–5 (2017)
42. J. Kang, B. Wang, Y. Zhang, J. An, Enhanced partially-parallel ldpc decoder for near earth applications, in *2021 IEEE 11th International Conference on Electronics Information and Emergency Communication (ICEIEC)2021 IEEE 11th International Conference on Electronics Information and Emergency Communication (ICEIEC)*, pp. 12–16 (2021)

Chapter 5
Secure Directional Modulation for LOS Communications

In this chapter, we introduce a multi-user hybrid beamforming design for a multi-user physical layer security modulation technique. Section 5.1 gives the background and motivation of the beamforming design for physical layer security communication. Section 5.2 introduces the system model of multi-user hybrid millimeter wave security communication. Section 5.3 presents the CE-based analog beamforming scheme. Section 5.4 discusses the CE-based multi-user hybrid beamforming scheme. Section 5.5 shows the simulation results. Section 5.6 presents the conclusion.

5.1 Introduction

With the explosive growth of commercial communication services, such as virtual reality and vision-based sensors [1–6], it is required to support a large number of low-latency and high-speed interconnection devices in 6G communication networks, which is difficult to achieve in low frequency bands. Therefore, millimeter wave (mmWave) band has become a natural choice to implement the high data rate transmission [7–10]. In addition, due to the high speed and frequent information exchange, each connected device also faces the problem of security communication [11]. The source encryption and decryption are mainly adopted by the existing security communication solutions, which have high complexity and huge resource consumption, resulting in a waste of resources in the entire system. The physical layer security (PLS) transmission technology uses the randomness of the wireless channel to achieve better secure transmission with low resource occupation [12–15].

The beamforming design for physical layer security communication has been studied by some literature recently. Specifically, W. Zhang et al. have proposed an artificial-noise-aided optimal beamforming scheme to minimize the transmit power at the base station, while meeting the minimum secrecy rate requirements of all secret messages [16]. The beamforming design scheme of optimal power allocation between

J. Li et al., *Key Technologies of High Frequency Wireless Communications*,
https://doi.org/10.1007/978-981-96-5894-7_5

the information-bearing signal and the artificial noise signal is introduced in [17] to enhance the physical layer security. J. Song et al. have proposed a beamforming scheme to maximize the secrecy rate of the desired user to enhance the physical layer security in dual-polarized millimeter wave systems [18]. A transmit zero forcing beamforming technique and the optimum power allocation algorithms have been studied in [19] to improve the physical layer security for multiple-input multiple-output non-orthogonal multiple access system. All the above literature has analyzed the system secure performance such as the secrecy rate and outage probability. In terms of the physical layer of security modulation technology, the deep nulls can be formed in the direction of the eavesdroppers to suppress the data reception and copying with the precise knowledge of channel state information [20]. However, the channel state of an eavesdropper is difficult to obtain because of the non-cooperation of eavesdroppers, so the deep null method is not practical. On the other hand, when the channel state of an eavesdropper is unknown, the transmitter sends data in the direction of the target user and the artificial noise in other directions [21–23]. Due to using part of the power to transmit noise, the power of the target signal is reduced, resulting in a decrease in the system's energy efficiency.

Physical layer security modulation techniques, such as directional modulation, which is shown in Table 5.1, transmit a defined constellation diagram in the direction of a target user, and transmit a noise pattern in other directions, which has been extensively studied in millimeter wave communication systems. The authors of [24] have proposed a low-complexity directional modulation technique, which can modulate the radiation pattern by selecting an antenna subset randomly for every symbol. The authors of [25] have introduced a switched phased-array transmission architecture, which consists of a conventional phased-array followed by antennas with a switch circuit to enhance physical layer security. Furthermore, a new scheme of optimized weight antenna subset transmission techniques based on programmable weight phased array architecture has been proposed in [26], which generates more artificial noise than conventional schemes in the undesired directions. Moreover, the authors of [27] have proposed a secure transmission algorithm by varying the transmitting weight vector at symbol rate based on the aid of polygon construction in the complex plane.

Due to the multiple transmission paths in the multi-user system, multiple security beams need to be generated at the same time to transmit data. However, the analog arrays used by the above mentioned secure transmission schemes can only be applied

Table 5.1 The comparison of physical layer modulation technology

References	Method	Scenario	Sidelobe suppression
[24]	Antenna subset modulation	Single user	5 dB
[25]	Switched antenna subset modulation	Single user	–
[26]	Programmable weight phased-array transmission	Single user	5.1 dB
[27]	A relaxed symbol region transmission	Single user	–
This	Hybrid directional modulation	Multi-user	10 dB

to a single target user scenario. Besides, it is hard to realize the fully-digital processing at millimeter wave frequencies with wide bandwidths and large antenna arrays because of the high cost and power consumption with mixed-signal components such as high-resolution analog to digital converters. Therefore, the hybrid beamforming scheme is adopted to service multiple users in this chapter.

To our best knowledge, we first propose the hybrid beamforming for multi-user physical layer security modulation techniques. In addition, we also propose cross-entropy to optimize the sidelobe energy, which makes the energy of an eavesdropper lower to further enhance the secure performance. Our simulation results demonstrate that the transmission structure can achieve a better multi-user physical layer security performance with lower sidelobe energy than the traditional schemes.

Notation: We use the following notations throughout the chapter. $(\cdot)^{\mathrm{T}}$, $(\cdot)^{\mathrm{H}}$ and $(\cdot)^{-1}$ denote the transpose, conjugate transpose, and inverse of a matrix or vector, respectively; We use the following variables throughout the chapter, which is shown in Table 5.2.

5.2 System Model

In this section, the multi-user hybrid millimeter wave security communication system is introduced, which is shown in Fig. 5.1. There is a base station Alice with multiple target users Bob_1, Bob_2,...,Bob_L and an eavesdropper Eve. The antenna subset is driven to transmit signal to ensure that the target user receives the modulated signal

Table 5.2 The main variables table

Variable	Description
N	Number of antennas
N_{RF}	Number of RF chains
N_{M}	Number of candidate combinations
N_{iter}	Number of iterations
N_c	Number of random combinations
M	Number of antennas with a weight of 1
L	Number of target users
k	Antenna weight
α	Smoothing parameter
$\mathbf{b}$	The weight of antennas
$\mathbf{y}$	The received signal of users
$\mathbf{x}$	The transmitted signal of users
$\mathbf{H}$	The channel matrix
$\mathbf{F}_{\mathrm{RF}}$	The analog precoding matrix
$\mathbf{F}_{\mathrm{BB}}$	The digital precoding matrix
$\mathbf{n}$	The noise vector

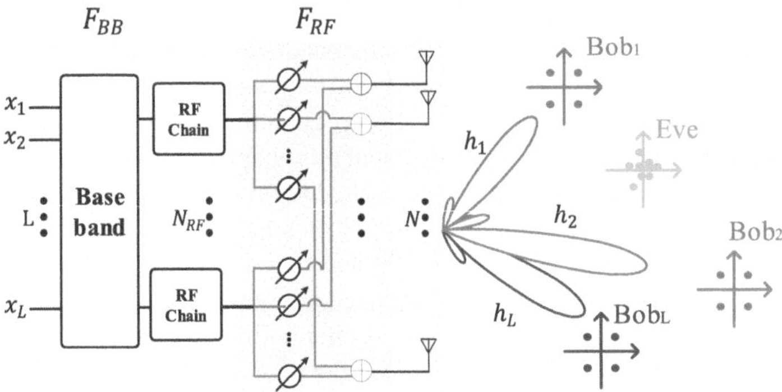

Fig. 5.1 Multi-user millimeter wave security communication system

correctly, while the eavesdropper receives the noise signal at the same time. The base station is equipped with N_{RF} radio frequency (RF) chains and N transmit antennas, which can service L target users and an eavesdropper. The system adopts the uniform linear array with the element space of one-half of the wavelength, which is also suitable for other antenna structures, such as uniform planar array or uniform spherical array.

The received signal of L users $\mathbf{y} = [y_1, y_2, \ldots, y_L]^H$ is expressed as:

$$\mathbf{y} = \mathbf{H}\mathbf{F}_{RF}\mathbf{F}_{BB}\mathbf{x} + \mathbf{n}, \tag{5.1}$$

where $\mathbf{H} = [\mathbf{h}_1, \mathbf{h}_2, \ldots, \mathbf{h}_L]^H \in \mathbb{C}^{L \times N}$ is a channel matrix with $\mathbf{h}_\ell, \ell = 1, 2, \ldots, L$, which is the $N \times 1$ channel vector between target user ℓ and transmitter. Besides, $\mathbf{F}_{RF} \in \mathbb{C}^{N \times N_{RF}}$ and $\mathbf{F}_{BB} \in \mathbb{C}^{N_{RF} \times L}$ are the analog precoding vector and digital precoding vector, and $\|\mathbf{F}_{RF}\mathbf{F}_{BB}\|^2 = \rho$, where ρ is the transmission power, and:

$$\begin{aligned} \mathbf{x} &= [x_1, x_2, \ldots, x_L]^H, \\ \mathbf{n} &= [n_1, n_2, \ldots, n_L]^H, \end{aligned} \tag{5.2}$$

where $x_\ell = \sqrt{E_s}e^{j\phi_M}$, $\mathbb{E}\left[|x_\ell|^2\right] = 1$, $\phi_M = \frac{2\pi m}{M}$, $m = 0, 1, \ldots, M - 1$ is the phase of M-PSK signal, $\sqrt{E_s}$ is the amplitude of signal, and $n_\ell \sim \mathcal{CN}\left(0, \frac{N_0}{2}\right), \ell = 1, 2, \ldots, L$.

Without loss of generality, we assume that the channel state information of the target user is perfectly known and the channel information of the eavesdropper is not known. In addition, the target user and the eavesdropper are not in the same direction. We only consider the LOS communication, due to the large channel attenuation of the millimeter wave [22, 28]. In order to simplify the mathematical model, we consider the two-dimensional x-y plane model, which only recognizes the azimuth angle and does not consider the influence of the elevation angle. The channel between transmitter and receiver at angle θ is:

$$\mathbf{h}_\ell(\theta_\ell) = \left[e^{-j\left(\frac{N-1}{2}\right)\frac{2\pi d\cos\theta_\ell}{\lambda}}, \; e^{-j\left(\frac{N-3}{2}\right)\frac{2\pi d\cos\theta_\ell}{\lambda}}, \; \ldots, \; e^{j\left(\frac{N-1}{2}\right)\frac{2\pi d\cos\theta_\ell}{\lambda}} \right]^H \ell = 1, \ldots, L, \quad (5.3)$$

where λ is the wavelength and d is the distance between the antennas. The L channels at angle $\theta_1, \theta_2, \ldots, \theta_L$ compose the channel matrix $\mathbf{H}$.

The hybrid beamforming scheme is used in the base station to generate multi-beams according to the direction angle of the target users. The base station first uses a secure analog beamforming scheme to generate analog beams in multiple desired directions, then uses minimum mean square error (MMSE) to design the digital beamforming matrix to eliminate inter-beam interference.

As for the analog security beamforming scheme, the choice of antenna weight affects the performance of security communication. The conventional, switched, reverse and programmable weighted phased-array transmission scheme is considered in [26] with different antenna weights, which is shown in Table 5.3. The analog beamforming vector $\mathbf{w}$ is denoted as the ℓ-th column of F_{RF}, which is shown as follow:

$$\mathbf{w} = \frac{1}{S} \left[\mathbf{b} \odot h\left(\theta_\ell\right) \right], \quad (5.4)$$

Table 5.3 The comparison of traditional beamforming scheme and security beamforming scheme

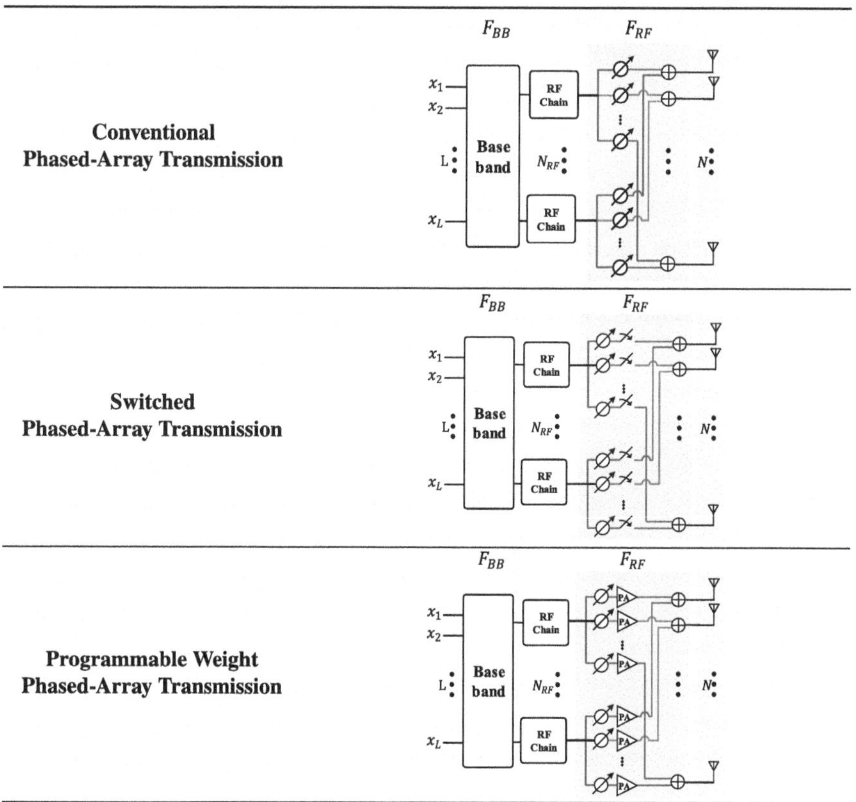

where $\mathbf{b} = [b_1, b_2, \ldots, b_N]^H$ is the weight of N antennas, $b_n \in \{k, 1\}, n = 1, \ldots, N$, and $\odot$ represents the dot product. We assume that in the antenna weight selection, $N - M$ antennas are set weight k, and else 1, then $S = M + (N - M)k$.

The received signal can be expressed as:

$$\begin{aligned} y_\ell &= h(\theta)\,\mathbf{w}_\ell x_\ell + n_\ell \\ &= \rho(\theta, \mathbf{b})x_\ell + n_\ell, \end{aligned} \tag{5.5}$$

where $\rho(\theta, \mathbf{b})$ is the scaling factor,

$$\rho(\theta, \mathbf{b}) = \frac{1}{S} \sum_{n=0}^{N-1} b_n e^{j\left(n - \frac{N-1}{2}\right)\left(\gamma_\theta - \gamma_{\theta_\ell}\right)}, \tag{5.6}$$

where $\gamma_\theta = \frac{2\pi d}{\lambda}\cos\theta$. Different weight vectors are generated to support different antenna subset transmission techniques by choosing different values of k [26].

In the traditional analog beamforming scheme, all the elements of antenna array have the same weight, that is, $b_n = k = 1$, $S = N$ at any discrete-time. The scaling factor $\rho(\theta)$ can be written as:

$$\begin{aligned} \rho(\theta, \mathbf{b}) &= \frac{1}{S} \sum_{n=0}^{N-1} b_n e^{j\left(n - \frac{N-1}{2}\right)\left(\gamma_\theta - \gamma_{\theta_\ell}\right)} \\ &= \frac{1}{N} \frac{\sin\left(\frac{N\left(\gamma_\theta - \gamma_{\theta_\ell}\right)}{2}\right)}{\sin\left(\frac{\left(\gamma_\theta - \gamma_{\theta_\ell}\right)}{2}\right)}. \end{aligned} \tag{5.7}$$

The constellation diagram of the target user is exactly the same as any other angle users, only the power is different. In addition, Eq. (5.7) shows that its radiation pattern is a Sinc wave, which is shown in Fig. 5.2. If the eavesdropper uses a large-aperture

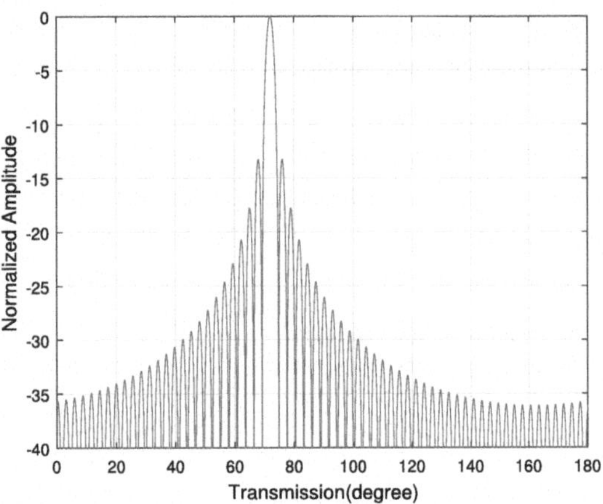

Fig. 5.2 The radiation pattern of the traditional analog beamforming scheme

antenna or ultra-low sensitivity receiving technology, it is possible for the eaves-dropper to recover the target user's signal. Therefore, the traditional beamforming scheme has poor security communication performance.

In the optimal weight antenna subset beamforming scheme, it controls the weight of the transmitted signal through a programmable amplifier, and chooses weight 1 with probability p, weight k with probability $1 - p$, so the antenna weight b obeys the Bernoulli distribution:

$$b_n \sim \text{Bern}(p) = \begin{cases} 1 \ \text{probability}(p) \\ k \ \text{probability}(1 - p). \end{cases} \tag{5.8}$$

According to [26], when N is large enough, the scaling factor ρ obeys a Gaussian distribution:

$$\mu(\theta) = \frac{1}{N} \frac{\sin\left(\frac{N(\gamma_\theta - \gamma_{\theta_\ell})}{2}\right)}{\sin\left(\frac{(\gamma_\theta - \gamma_{\theta_\ell})}{2}\right)},$$

$$\sigma^2(\theta) = \frac{(k-1)^2 p(1-p)}{N(p + k(1-p))^2}. \tag{5.9}$$

The optimal weight antenna subset beamforming scheme maximizes the artificial noise in the direction of an eavesdropper to optimize the weight of each antenna transmission. According to [26], when $k = \frac{1-M}{N-M}$, the antenna subset achieves an optimal secure performance.

The constellation diagram of a different phased-array transmission scheme in the direction of the eavesdropper is shown as Fig. 5.3. It can be seen from the figure

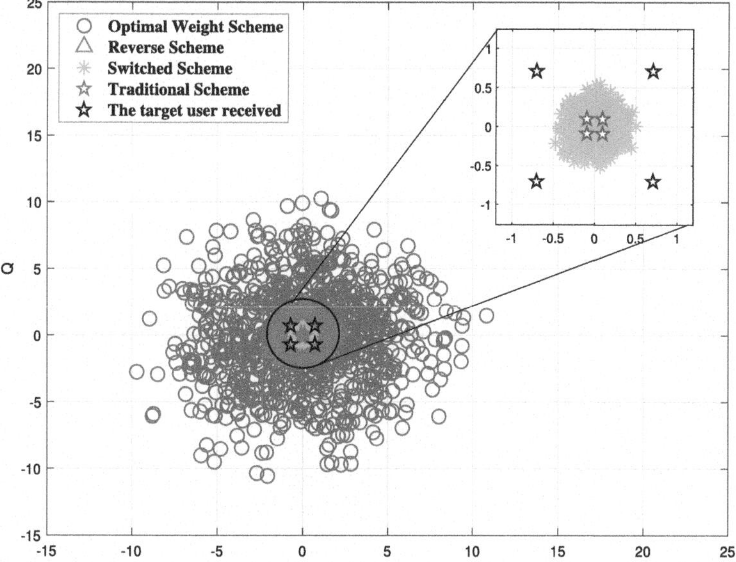

Fig. 5.3 The constellation diagram in the direction of the eavesdropper

that only the traditional beamforming scheme has the same constellation diagram between the target user and the eavesdropper, and the other schemes' constellation diagrams of the eavesdropper are all artificial noise because of the selection of a subset of antennas randomly. The optimal weighted antenna subset beamforming scheme has the highest degree of dispersion and the best physical layer security performance, which is adopted to implement analog beamforming in this chapter.

5.3 CE-Based Analog Precoding Scheme

The above security communication beamforming schemes select appropriate antenna combinations to create artificial noise at the eavesdropper's direction, but the sidelobe energy of these directions is too high. In order to reduce the signal energy received by the eavesdropper, we use the cross-entropy iterative algorithm to select antenna combinations with low sidelobe energy, so as to ensure that the eavesdropper receives lower signal energy to further enhance the secure performance.

Multi-user hybrid beamforming scheme first uses analog beamforming to generate multiple beams, and then uses MMSE methods to eliminate inter-beam interference. We first consider the analog beamforming. We set b as antenna combination, and use a cross-entropy iterative algorithm to choose optimal antenna combination b^*:

$$b^* = \underset{b \in \mathcal{B}}{\arg\min} \left(\text{Fitness} < T_H\right), \qquad (5.10)$$

where Fitness is the optimized objective function, T_H is the threshold and $\mathcal{B}$ is the set of all antenna combinations. The probability density distribution function of the antenna combination can be defined as:

$$f(b, p) = \prod_{i=1}^{N} p_i^{I_i(b)} (1 - p_i)^{1 - I_i(b)}, \qquad (5.11)$$

where N is the number of transmitting antennas, p_i is the probability that the antenna i is selected, $I_i(b)$ is the indicator function. For a given probability distribution p, the possibility of $Fitness < T_H$ is:

$$\varphi(T_H) = P\left(\text{Fitness} < T_H\right) = \sum_{b \in \Omega} I\left(\text{Fitness} < T_H\right) f(b, p), \qquad (5.12)$$

and the indicator function $I_i(b)$ is:

$$I\left(\text{Fitness} < T_H\right) = \begin{cases} 1 & \text{Fitness} < T_H \\ 0 & \text{Fitness} \geq T_H \end{cases}. \qquad (5.13)$$

In order to obtain the optimal probability density distribution function, we set $g(b)$ as another probability density distribution function about b, then φ can be written as:

$$\varphi(T_H) = \int \frac{I\,(\text{Fitness} < T_H) f(b, p)}{g(b)} g(b)\mathrm{db}$$
$$= \mathbb{E}\left[\frac{I\,(\text{Fitness} < T_H) f(b, p)}{g(b)}\right]. \tag{5.14}$$

According to [29], the optimal probability density distribution function $g^*(b)$ is:

$$g^*(b) = \frac{I\,(\text{Fitness} < T_H) f(b, p)}{\varphi}. \tag{5.15}$$

The purpose of the cross-entropy iterative algorithm is to minimize the K-L divergence between $f(b, p)$ and the optimal probability density distribution function $g^*(b)$. The K-L divergence can be written as:

$$D(g, h) = \mathbb{E}\left[\ln \frac{g(x)}{h(x)}\right] = \int g(x) \ln g(x)\mathrm{dx} - \int g(x) \ln h(x)\mathrm{dx}. \tag{5.16}$$

Since $g^*(b)$ is not related to the p, minimizing K-L divergence is equivalent to solving the following maximization problem:

$$\max_p \int_\Omega g^*(b) \ln f(b, p)\mathrm{db}. \tag{5.17}$$

Substitute formula (5.15) into formula (5.17),

$$\max_p \int_\Omega \frac{I\,(\text{Fitness} < T_H) f(b, p)}{\varphi} \ln f(b, p)\mathrm{db}. \tag{5.18}$$

The maximization problem can be reduced to:

$$\max_p \frac{1}{N} \sum_{n=1}^{N} I\,(\text{Fitness}^{(n)} < T_H) \ln f(b^{(n)}, p)$$
$$= \max_p \frac{1}{N} \sum_{n=1}^{N} I\,(\text{Fitness}^{(n)} < T_H) \ln \prod_{i=1}^{N} p_i^{I_i(b^{(n)})} (1 - p_i)^{1 - I_i(b^{(n)})}. \tag{5.19}$$

This function is a convex function, and its partial derivative can be 0 to find the optimal solution:

$$\frac{\partial}{\partial p_i} \sum_{n=1}^{N} I\,(\text{Fitness}^{(n)} < T_H) \ln \prod_{i=1}^{N} p_i^{I_i(b^{(n)})} (1 - p_i)^{1 - I_i(b^{(n)})}$$
$$= \sum_{n=1}^{N} I\,(\text{Fitness}^{(n)} < T_H) \frac{\partial}{\partial p_i} \ln \prod_{i=1}^{N} p_i^{I_i(b^{(n)})} (1 - p_i)^{1 - I_i(b^{(n)})} \tag{5.20}$$
$$= \sum_{n=1}^{N} I\,(\text{Fitness}^{(n)} < T_H) \frac{I_i(b^{(n)}) - p_i}{p_i(1 - p_i)}.$$

We set formula (5.20) equal to 0, the probability of antenna combination iteration can be obtained as:

$$p_i = \frac{\sum_{n=1}^{N} I\left(\text{Fitness}^{(n)} < T_H\right) I_i\left(b^{(n)}\right)}{\sum_{n=1}^{N} I\left(\text{Fitness}^{(n)} < T_H\right)} \quad i = 1, 2, \ldots, N. \tag{5.21}$$

According to the iterative rule in formula (5.21), the antenna combination selection strategy can be continuously updated until it converges to the optimal solution. In each iteration process, new samples are generated according to the probability distribution of the previous iteration. These samples have a greater probability of meeting the judgment conditions of the objective function, and then the probability of the next iteration is calculated based on the samples generated this time. Until the stopping strategy is met, the cross-entropy algorithm terminates iteration.

The following is a simulation of the cross-entropy iterative algorithm. The relationship between the number of iterations and the objective function is simulated firstly. Figure 5.4 shows that, when the number of candidate signals is small, the objective function can converge quickly, but it may converge to the local maximum. In addition, the number of iterations required needs to increase to converge to the optimal value with the number of candidate signals increasing.

Figure 5.5 shows the relationship between the number of random combination and objective functions. As the number of random combinations in each iteration increases, the range of each traversal is larger, and it is easier to converge to the optimal value. The optimal value tends to stabilize with the number of random combinations increasing to 200. When the number of candidate combinations is 1/5 of the number of random combinations, it converges to the optimal value.

It can be seen that, as the number of iterations and combinations increase, the objective function can converge to the optimal value. In [30], the authors have proved that when selecting a sufficiently small value of α, the limiting probability of generating an optimal solution can be made arbitrarily close to 1. In addition, a smaller

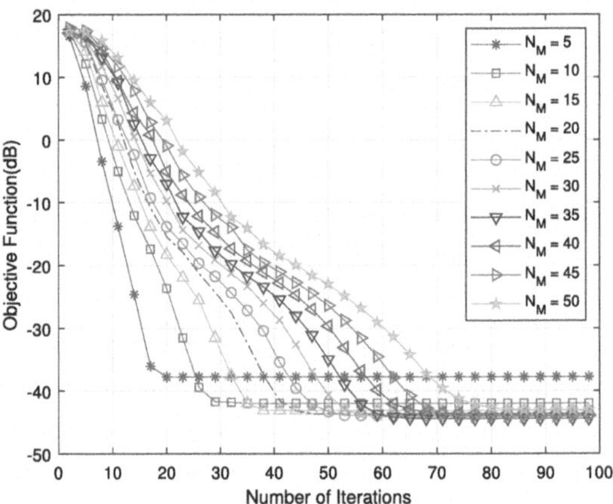

Fig. 5.4 The relationship between the number of iterations and the objective function

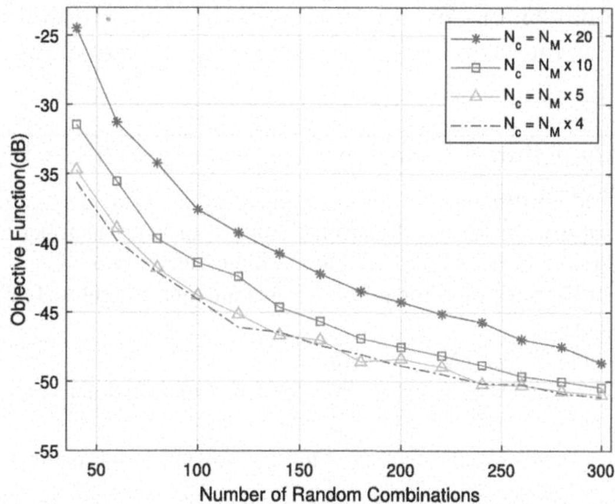

Fig. 5.5 The relationship between random combination number and objective function

value of α can reduce the empirical rate of convergence. There is a balance between an optimal solution with high probability and a fast rate of convergence. In order to reduce the computational complexity and increase the convergence rate, we set the smoothing parameter α to 1 in this chapter.

According to whether the transmitter knows the channel information of the eavesdropper, we have two optimization goals: one is to optimize the sidelobes of a single direction,

$$\text{Fitness} = |\rho(\theta, \mathbf{b})|; \tag{5.22}$$

the other is to optimize the sidelobes of all non-target directions,

$$\text{Fitness} = \max\left(|\rho(\theta_1, \mathbf{b})|, |\rho(\theta_2, \mathbf{b})|, \ldots, |\rho(\theta_N, \mathbf{b})|\right). \tag{5.23}$$

Then, the antenna subset is

$$\mathbf{b}^* = \operatorname*{argmax}_{\theta \in \Omega, \mathbf{b} \in \mathcal{B}} \text{Fitness} < T_H, \tag{5.24}$$

where Ω is the set of all non-target users, $\mathcal{B}$ is the set of all antenna combinations, θ is the direction angle of the eavesdropper, and T_H is the threshold value, which is used to control the energy of the antenna combination sidelobe.

In order to reduce the energy of the transmitter pattern sidelobe, we propose an analog beamforming scheme based on cross-entropy iterative algorithm which is shown in Algorithm 5.1. The algorithm first inputs the number of random antenna combinations N_C, the number of candidate signals N_M, the total number of iterations N_{iter}, and smoothing parameter α and initializes the probability of each antenna weight to 1 with $\mathbf{P}_b^0$ to 0.5. Then, the objective function value of each combination is calculated and sorted. Next, we take the first N_M sorted antenna combinations with $\mathbf{C}_b^{seq,1}$, $\mathbf{C}_b^{seq,2}$, ..., $\mathbf{C}_b^{seq,N_M}$ and calculate the iterative probability of the antenna combination. We use the probability to perform the next iteration until the number of iterations is completed, at last the algorithm outputs the antenna combination **b**.

Algorithm 5.1 Analog Beamforming Scheme Based on Cross-Entropy Iterative Algorithm.

Input: Number of Random Combinations N_C, Number of Candidate Combinations N_M, Total Number of Iterations N_{iter}, Smoothing Parameter α

Initialization: Probability of each antenna weight to 1, $\mathbf{P}_b^0 = 0.5 * \mathbf{1}_{1 \times N}$, Number of iterations $i = 0$

1: **for** $0 \leq i \leq N_{iter} - 1$ **do**

2:	Randomly generate N_C antenna combinations vector $\{\mathbf{b}_n\}_{n=1}^{N_c}$ with probability $\mathbf{P}_b^i$, $\mathbf{b}_n \in \{k, 1\}^{1 \times N}$

3:	Calculate the Fitness of the N_C antenna combinations as $\eta_1^i, \eta_2^i, \ldots, \eta_{N_c}^i$

4:	Sort η_n^i and get the sorted sequence as: $\eta_{seq,1}^i \leq \eta_{seq,2}^i \leq, \ldots, \leq \eta_{seq,N_c}^i$

5:	Take the first N_M antenna combination to calculate the iteration probability: $\mathbf{P}_b^{i+1} = (1 - \alpha)\mathbf{P}_b^i + \alpha \frac{1}{N_M} \left(\mathbf{C}_b^{seq,1} + \mathbf{C}_b^{seq,2} +, \ldots, + \mathbf{C}_b^{seq,N_M} \right)$

6:	$i = i + 1$.

7: **end for**

Output: Antenna Combination b.

Figure 5.6 shows the sidelobe energy of the emission pattern when the eavesdropper angle is 80°, the target user angle is 36° and the number of antennas is 128. The random selection algorithm is introduced in [31]. The simulation shows that the sidelobe energy suppression in the eavesdropper direction is 44.94 dB. Figure 5.7 shows the sidelobe energy of the emission pattern when the direction of the eavesdropper is unknown, the target user angle is 40 degrees and the number of antennas is 128. The simulation shows that the sidelobe energy of the CE optimal algorithm is 10 dB lower than the random selection algorithm.

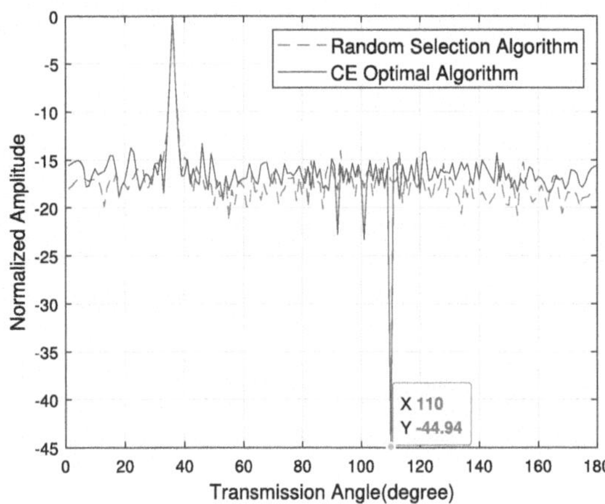

Fig. 5.6 The sidelobe energy of the emission pattern with the known channel state information of the eavesdropper

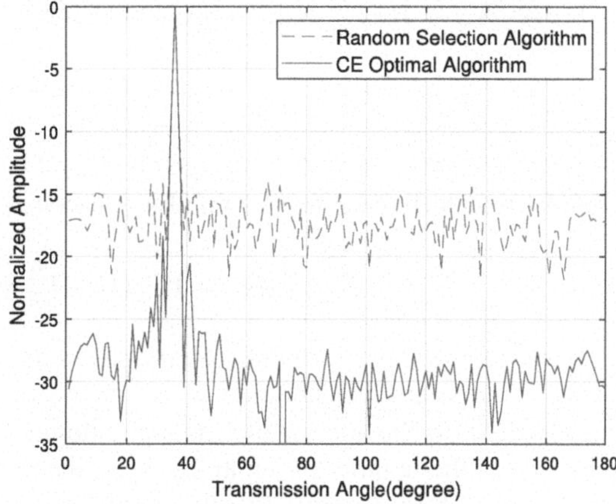

Fig. 5.7 The sidelobe energy of the emission pattern with the unknown channel state information of the eavesdropper

---

Algorithm 5.2 Hybrid Beamforming Scheme Based on Cross-Entropy Iterative Algorithm.

Input: Number of Random Combinations N_C, Number of candidate Combinations N_M, Total Number of iterations N_{iter}, Number of Users L, Smoothing Parameter α

Initialization: Probability of each antenna weight to 1, $\mathbf{P}_b^0 = 0.5 * \mathbf{1}_{1 \times N}$, Number of iterations $i = 0$

1: **for** $0 \leq i \leq N_{iter} - 1$ **do**

2:     Randomly generate N_C antenna combinations vector $\{\mathbf{b}_n\}_{n=1}^{N_c}$ with probability $\mathbf{P}_b^i$, $\mathbf{b}_n \in \{k, 1\}^{1 \times N}$

3:     Calculate the analog beamforming vector of the N_C antenna combination $\{\mathbf{F}_{RF_b}\}_{n=1}^{N_c}$

4:     Calculate the equivalent channel $\{\mathbf{H}_{eq}\}_{n=1}^{N_c}$ according to $\{\mathbf{F}_{RF_b}\}_{n=1}^{N_c}$

5:     Use MMSE algorithm to calculate the digital beamforming vector: $\{\mathbf{F}_{BB\_MMSE}\}_{n=1}^{N_c}$

6:     Calculate the Fitness of the N_C group antenna combinations as $\eta_1^i, \eta_2^i, \ldots, \eta_{N_c}^i$

7:     Sort η_n^i and get the sorted sequence as: $\eta_{seq,1}^i \leq \eta_{seq,2}^i \leq, \ldots, \leq \eta_{seq,N_c}^i$

8:     Take the first N_M antenna combination to calculate the iteration probability: $\mathbf{P}_b^{i+1} = (1 - \alpha)\mathbf{P}_b^i + \alpha \frac{1}{N_M} \left(\mathbf{C}_b^{seq,1} + \mathbf{C}_b^{seq,2} +, \ldots, + \mathbf{C}_b^{seq,N_M} \right)$

9:     $i = i + 1$.

10: **end for**

Output: Antenna Combination b.

5.4 CE-Based Multi-user Hybrid Precoding Scheme

The previous section has introduced the design of analog beamforming algorithm, this section mainly introduces the design of the multi-user hybrid beamforming algorithm. The received signal of the multi-user hybrid beamforming in security communication system is:

$$\mathbf{y} = \mathbf{H}\mathbf{F}_{RF_b}\mathbf{F}_{BB}\mathbf{x} + \mathbf{n}, \tag{5.25}$$

where $\mathbf{F}_{RF_b} = \mathbf{b} \odot \mathbf{F}_{RF}$ represents the dot product of $\mathbf{b}$ and each column of $\mathbf{F}_{RF}$.

We set $\mathbf{H}_{eq} = \mathbf{H}\mathbf{F}_{RF_b}$ as the equivalent channel, formula (5.25) can be written as:

$$\mathbf{y} = \mathbf{H}_{eq}\mathbf{F}_{BB}\mathbf{x} + \mathbf{n}. \tag{5.26}$$

We use MMSE methods to design the digital beamforming matrix $\mathbf{F}_{BB}$, thereby eliminating interference between multiple users.

$$\mathbf{F}_{BB\_MMSE} = \mathbf{H}_{eq}^{H}\left(\mathbf{H}_{eq}\mathbf{H}_{eq}^{H} + \beta I\right)^{-1}. \tag{5.27}$$

Based on the above formulas, We propose a multi-user hybrid procoding scheme based on cross-entropy iteration which is shown in Algorithm 5.2. The input of this algorithm is the number of random combinations N_C, the number of candidate signals N_M, the number of iterations N_{iter}, the number of users L, and smoothing parameter α. Then we initialize the probability of each antenna weight to 1 with $\mathbf{P}_b^0$ to 0.5. First, we generate N_C antenna combination vectors. According to the multi-user channel, we calculate the simulated precoding vector of N_C antenna combinations $\left\{\mathbf{F}_{RF_b}\right\}_{n=1}^{N_c}$, and the equivalent channel based on the simulated precoding vector $\left\{\mathbf{H}_{eq}\right\}_{n=1}^{N_c}$. Then, the MMSE algorithm is adopted to calculate the digital precoding vector to eliminate the interference between multiple user beams, so as to ensure the normal transmission of data between multiple users. Next, we calculate the fitness of the pattern sidelobes generated by the N_C antenna combinations and sort to obtain $\eta_{seq,1}^{i} \leq \eta_{seq,2}^{i} \leq, \ldots, \leq \eta_{seq,N_c}^{i}$. According to the sorted sequence, the first N_M antenna combination with $\mathbf{C}_b^{seq,1}, \mathbf{C}_b^{seq,2}, ..., \mathbf{C}_b^{seq,N_M}$ is used to calculate the probability $\mathbf{P}_b^{i+1}$ for the next iteration. Until the iteration ends, the antenna combination $\mathbf{b}$ with low sidelobes of the pattern is output. In addition, Algorithm 2 can also be applied to other channel models because it only relates to the channel matrix $\mathbf{H}$. However, due to the existence of multipath in other channel model, the sidelobe energy will increase, making the performance of the proposed algorithm worse. Besides, how to set threshold T_H for the fitness evaluation is an open question. We know that, when the threshold is smaller, the sidelobes of the pattern are lower but the number of antenna combinations that meet the requirements is less. Conversely, the threshold is larger, the sidelobes of the pattern are higher, but the number of antenna combinations that meets the requirements is greater.

Figure 5.8 shows the sidelobe energy of the emission pattern under the known eavesdropper angle. A multi-user hybrid beamforming algorithm based on cross-entropy is used to iterate the random antenna combination, and the sidelobe energy is selected to be lower. In the direction of the target user, the normalized amplitude of the mainlobe energy of the target user is all 0 dB, and the sidelobe suppression reaches 46.45 dB in the direction of the eavesdropper.

Figure 5.9 shows the sidelobe energy of the emission pattern at the unknown eavesdropper angle. By optimizing the sidelobe energy in the direction of non-target users, the antenna combination with lower sidelobe energy is obtained. The sidelobe energy with the CE optimal algorithm is 10 dB lower than the random selection algorithm, which allows multiple target users to communicate normally while eaves-

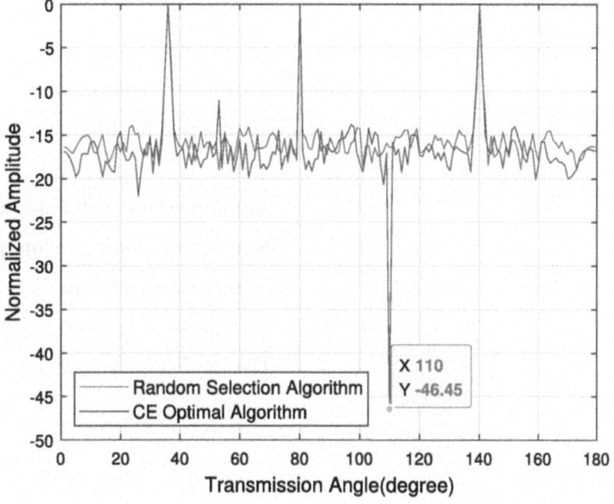

Fig. 5.8 The sidelobe energy of the emission pattern with the known channel state information of the eavesdropper

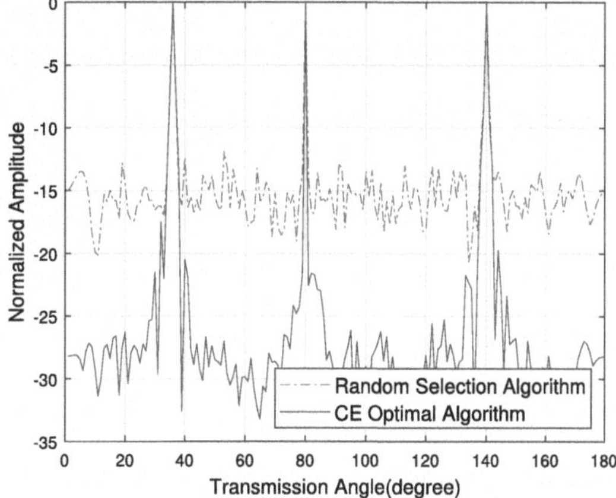

Fig. 5.9 The sidelobe energy of the emission pattern with the unknown channel state information of the eavesdropper

droppers receive low energy noise constellations, thereby further improving security communication.

5.5 Simulation Results

According to the CE-based multi-user hybrid beamforming algorithm, the physical layer security performance of the millimeter-wave communications system is simulated in this section. The influence of using the CE optimal algorithm and the random selection algorithm on the sidelobe energy and symbol error rate of the system is analyzed. The simulation parameters are shown in Table 5.4.

Figure 5.10 shows the probability density distribution function of the sidelobe energy. This figure first calculates the emission pattern based on the CE optimal

Table 5.4 Simulation parameter table

Parameter	Value
Modulation	QPSK
Number of antennas	128
Number of RF chains	3
Number of target users	3
Number of random antenna weight vectors	100
Number of candidate antenna weight vectors	25
Number of iterations	40
The angle of target users (degrees)	36, 80, 140
The angle of Eve (degrees)	110

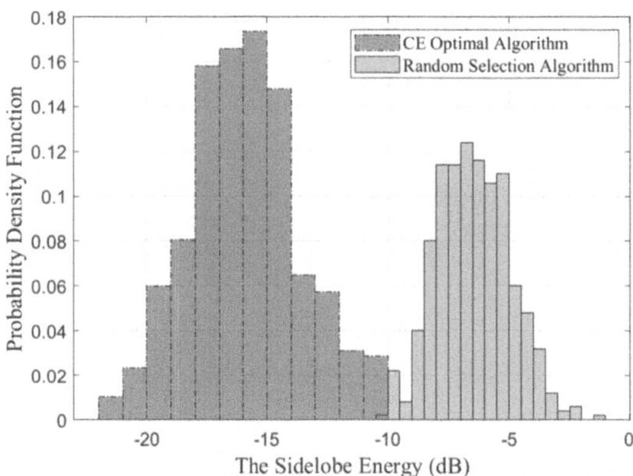

Fig. 5.10 The probability density distribution function of the sidelobe energy

algorithm and the random selection algorithm and normalizes the sidelobe energy by subtracting the mainlobe energy. Then, the normalized maximum sidelobe is obtained. The probability density distribution function of the maximum sidelobe energy is acquired by Monte Carlo simulation. It can be shown that the sidelobe energy obtained by the random selection algorithm is mainly distributed around -7 dB, while the sidelobe energy of the CE optimal algorithm is -17 dB, and both have a Gaussian-like distribution.

Figure 5.11 shows the relationship between the angle of the eavesdropper and the symbol error rate. The eavesdropper demodulates the symbols of target user1, target user2, target user3 from different angles and calculates the symbol error rate. It can be seen that, when the eavesdropper and the target user are located in the same direction, the symbol error rate tends to zeros. The symbol can be perfectly demodulated when the eavesdropper has a large-scale antenna or an ultra-low sensitivity receiver. Furthermore, the symbol error rate is about 0.75 when the eavesdropper is at other angles. Due to the constellation diagram being noise, it is difficult for the eavesdropper to demodulate the symbols.

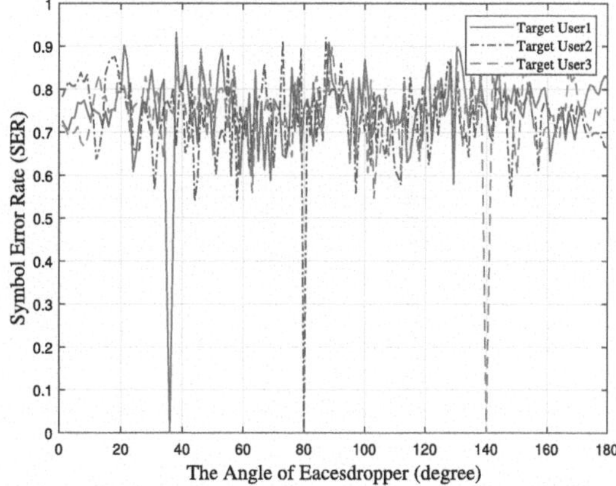

Fig. 5.11 The symbol error rate against the angle of the eavesdropper

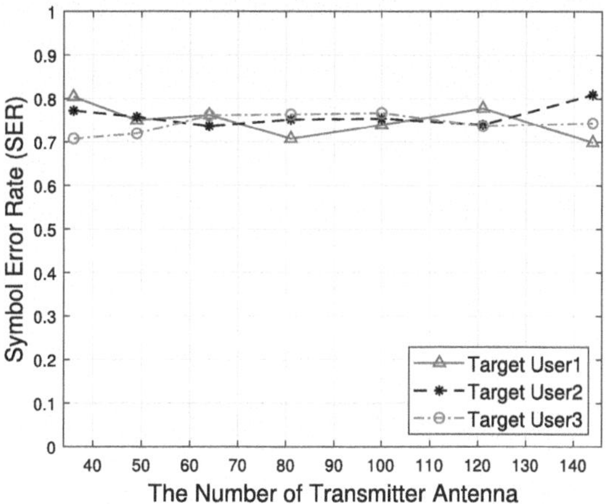

Fig. 5.12 The symbol error rate against the number of the transmitter antenna

Figure 5.12 shows the relationship between the number of transmitter antennas and the symbol error rate. As the number of transmitter antennas increases, the symbol error rate is always approximately 0.75. The eavesdropper cannot correctly obtain the information transmitted by the target users.

Figure 5.13 shows the relationship between the SNR of the eavesdropper and the symbol error rate. When the direction of the eavesdropper is not the same as the target user, as the SNR increases, the symbol error rate is always approximate 0.75. The signal cannot be demodulated correctly, which has a good security communication performance.

Figure 5.14 shows the relationship between the number of target users and the symbol error rate. As the number of target users increases, the symbol error rate does not deteriorate and is maintained at about 0.75 when the number of target users is less than that of the RF chains. Therefore, the CE optimal algorithm can achieve a better physical layer security communication performance.

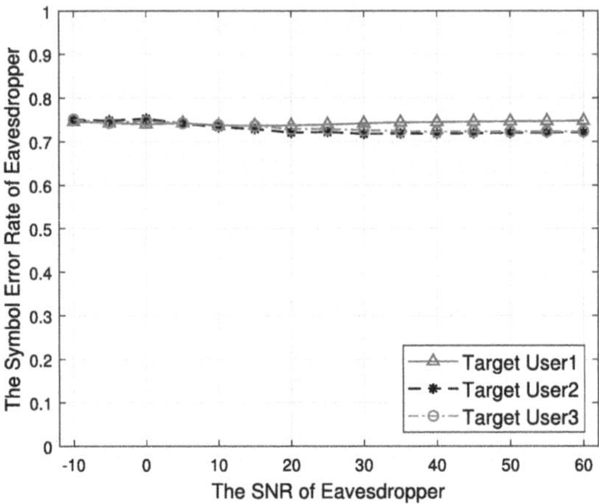

Fig. 5.13 The symbol error rate against the SNR of the eavesdropper

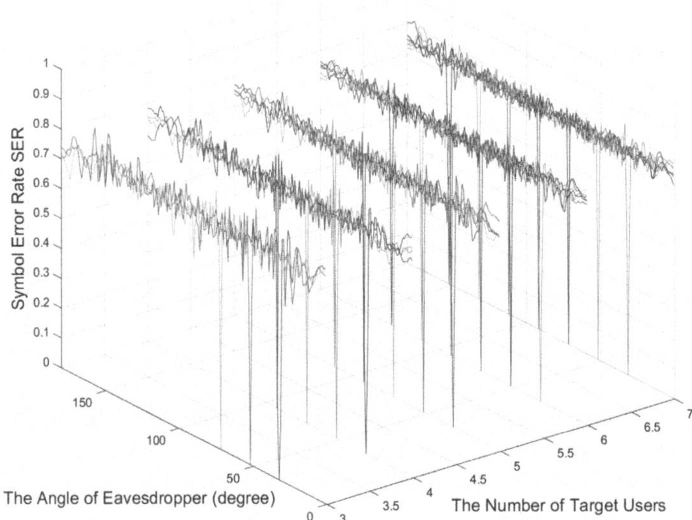

Fig. 5.14 The relationship between the number of target users and the symbol error rate

5.6 Conclusions

In this chapter, we propose a CE optimal based hybrid beamforming design algorithm for multi-user physical layer security modulation techniques. We first adopt the hybrid beamforming method and randomly select antennas to transmit the signal at the symbol rate. In addition, in order to solve the problem of the high sidelobe energy of randomly selected antennas, this chapter proposes a cross-entropy iteration method to choose the optimal antenna combination to transmit signals, thereby reducing the energy of eavesdropper-received signals and further improving the system's security performance. The simulation shows that the proposed method has about 10 dB lower sidelobe energy than does the random selection method. Besides, while multiple target users are communicating normally, the eavesdropper's symbol error rate of QPSK is always 0.75, which has a good physical layer security communication performance. In future research, a hybrid beamforming design scheme for physical layer secured mmWave communications in other channel models will be studied to realize the multi-user physical layer security.

References

1. C. Psomas, I. Krikidis, Energy beamforming in wireless powered mmWave sensor networks. IEEE J. Sel. Areas Commun. **37**(2), 424–438 (2019)
2. B. Chang, G. Zhao, L. Zhang, M.A. Imran, Z. Chen, L. Li, Dynamic communication QoS design for real-time wireless control systems. IEEE Sensors J. **20**(6), 3005–3015 (2020)
3. F. Guo, F.R. Yu, H. Zhang, X. Li, H. Ji, V.C.M. Leung, Enabling massive IoT toward 6G: a comprehensive survey. IEEE Internet Things J. **8**(15), 11891–11915 (2021)
4. W. Lin, T. Matsumoto, Performance analysis of distortion-acceptable cooperative communications in wireless sensor networks for internet of things. IEEE Sensors J. **19**(5), 1979–1989 (2019)
5. S.A. Busari, K.M.S. Huq, S. Mumtaz, L. Dai, J. Rodriguez, Millimeter-wave massive MIMO communication for future wireless systems: a survey. IEEE Commun. Surveys Tutorials **20**(2), 836–869 Second quarter (2018)
6. X. Gao, S. Feng, D. Niyato, P. Wang, K. Yang, Y.-C. Liang, Dynamic access point and service selection in backscatter-assisted RF-powered cognitive networks. IEEE Internet Things J. **6**(5), 8270–8283 (2019)
7. J. Zhang, Y. Huang, J. Wang, R. Schober, L. Yang, Power-efficient beam designs for millimeter wave communication systems. IEEE Trans. Wirel. Commun. **19**(2), 1265–1279 (2020)
8. Y. Ju, H.-M. Wang, T.-X. Zheng, Q. Yin, M.H. Lee, Safeguarding millimeter wave communications against randomly located eavesdroppers. IEEE Trans. Wirel. Commun. **17**(4), 2675–2689 (2018)
9. R. Liu, G. Yu, J. Yuan, G.Y. Li, Resource management for millimeter-wave ultra-reliable and low-latency communications. IEEE Trans. Commun. **69**(2), 1094–1108 (2021)
10. J. An, Y. Zhang, X. Gao, K. Yang, Energy-efficient base station association and beamforming for multi-cell multiuser systems. IEEE Trans. Wirel. Commun. **19**(4), 2841–2854 (2020)
11. H.M. Furqan, M. Solaija, H. Türkmen, H. Arslan, Wireless communication, sensing, and REM: a security perspective. IEEE Open J. Commun. Soc. **2**, 287–321 (2021)
12. Y. Wu, A. Khisti, C. Xiao, G. Caire, K.-K. Wong, X. Gao, A survey of physical layer security techniques for 5G wireless networks and challenges ahead. IEEE J. Sel. Areas Commun. **36**(4), 679–695 (2018)
13. W.-Q. Wang, Z. Zheng, Hybrid MIMO and phased-array directional modulation for physical layer security in mmWave wireless communications. IEEE J. Sel. Areas Commun. **36**(7), 1383–1396 (2018)
14. L. Qing, H. Guangyao, F. Xiaomei, Physical layer security in multi-hop AF relay network based on compressed sensing. IEEE Commun. Lett. **22**(9), 1882–1885 (2018)
15. F. Jameel, S. Wyne, G. Kaddoum, T.Q. Duong, A comprehensive survey on cooperative relaying and jamming strategies for physical layer security. IEEE Commun. Surveys Tutorials **21**(3), 2734–2771, Third quarter (2019)

16. W. Zhang, J. Chen, Y. Kuo, Y. Zhou, Transmit beamforming for layered physical layer security. IEEE Trans. Veh. Technol. **68**(10), 9747–9760 (2019)

17. D. Hu, P. Mu, W. Zhang, W. Wang, Minimization of secrecy outage probability with artificial-noise-aided beamforming for MISO wiretap channels. IEEE Commun. Lett. **24**(2), 401–404 (2020)

18. B. Song, B. Lee, J. Park, M.-S. Lee, J.-H. Lee, Beamformer design for physical layer security in dual-polarized millimeter wave channels. IEEE Trans. Veh. Technol. **69**(10), 12306–12311 (2020)

19. N. Nandan, S. Majhi, H.-C. Wu, Beamforming and power optimization for physical layer security of MIMO-NOMA based CRN over imperfect CSI. IEEE Trans. Veh. Technol. **70**(6), 5990–6001 (2021)

20. T. Lv, H. Gao, S. Yang, Secrecy transmit beamforming for heterogeneous networks. IEEE J. Sel. Areas Commun. **33**(6), 1154–1170 (2015)

21. Y. Gu, Z. Wu, Z. Yin, X. Zhang, The secrecy capacity optimization artificial noise: a new type of artificial noise for secure communication in MIMO system. IEEE Access **7**, 58353–58360 (2019)

22. T. Xie, J. Zhu, Y. Li, Artificial-noise-aided zero-forcing synthesis approach for secure multi-beam directional modulation. IEEE Commun. Lett. **22**(2), 276–279 (2018)

23. X. Yu, Y. Hu, Q. Pan, X. Dang, N. Li, M.H. Shan, Secrecy performance analysis of artificial-noise-aided spatial modulation in the presence of imperfect CSI. IEEE Access **6**, 41060–41067 (2018)

24. N. Valliappan, A. Lozano, R.W. Heath, Antenna subset modulation for secure millimeter-wave wireless communication. IEEE Trans. Commun. **61**(8), 3231–3245 (2013)

25. N.N. Alotaibi, K.A. Hamdi, Switched phased-array transmission architecture for secure millimeter-wave wireless communication. IEEE Trans. Commun. **64**(3), 1303–1312 (2016)

26. Y. Hong, X. Jing, H. Gao, Programmable weight phased-array transmission for secure millimeter-wave wireless communications. IEEE J. Sel. Topics Signal Process. **12**(2), 399–413 (2018)

27. X. Zhang, X.-G. Xia, Z. He, X. Zhang, Phased-array transmission for secure mmWave wireless communication via polygon construction. IEEE Trans. Signal Process. **68**, 327–342 (2020)

28. M. Hafez, M. Yusuf, T. Khattab, T. Elfouly, H. Arslan, Secure spatial multiple access using directional modulation. IEEE Trans. Wirel. Commun. **17**(1), 563–573 (2018)

29. R.Y. Rubinstein, D.P. Kroese, *The cross-entropy method: a unified approach to combinatorial optimization, Monte-Carlo simulation and machine learning; Berlin* (Springer, Germany, 2013)

30. A. Costa, O.D. Jones, D. Kroese, Convergence properties of the cross-entropy method for discrete optimization. Oper. Res. Lett. **35**, 573–580 (2007)

31. M. Hafez, H. Arslan, On directional modulation: An analysis of transmission scheme with multiple directions, in *Proceedings of the 2015 IEEE International Conference on Communication Workshop (ICCW)*, London, UK, pp. 459–463 (2015)

Chapter 6
Secure Directional Modulation in RIS-Aided Networks

In this chapter, we elaborate on secure directional modulation in RIS-aided networks with a low-sidelobe hybrid beamforming approach. Section 6.1 introduces the background and motivation of RIS-aided wireless secure communication system. Section 6.2 introduces the system model of RIS-aided communication networks. Section 6.3 presents the cross-entropy method based hybrid beamforming scheme. Section 6.4 gives the simulation results, and Sect. 6.5 summarizes the chapter.

6.1 Introduction

With the ability to intelligently manipulate the propagation channels, RIS, a composition of massive reconfigurable passive reflection units on a planar surface, has been shown to significantly enhance the spectrum efficiency and coverage capability of next generation wireless communication networks [1–3]. In the meantime, the physical-layer security issue has become one prominent research topic in RIS-aided communication networks [4–6].

RIS can provide a new dimension for the secure beamforming design of the mmWave communication networks [7–9]. Specifically, the authors in [7] have considered an analytical expression of the ergodic secrecy rate with large dimensional random matrix theory of RIS-aided wireless secure communication system. In [8], analytical approximation has been performed to estimate the outage probability, achievable rate and symbol error rate in RIS-aided communication networks. In the presence of discrete phase shifts, the secrecy outage probability and achievable rate of RIS-aided secure communication have been discussed in [9]. The above-mentioned researches mainly focus on the optimization of the secure performance such as the secrecy rate and outage probability which requires a weaker channel of the eavesdropper than the target user. However, little attention has been paid to promoting the secrecy using physical-layer techniques in RIS-aided communication networks.

J. Li et al., *Key Technologies of High Frequency Wireless Communications*,
https://doi.org/10.1007/978-981-96-5894-7_6

In this chapter, we propose to incorporate the secure directional modulation in RIS-aided communication networks to assure physical-layer security at undesired directions. As a promising physical-layer technique, the secure directional modulation can generate randomized signals for eavesdroppers by elaborately selecting the antenna subsets, and has been widely studied in mmWave communication networks. A low complexity direction modulation scheme has been proposed in [10] to facilitate mmWave physical-layer security communication for projecting a defined constellation in the desired direction and expanding randomized constellation in other directions. Later on, a switched phased array transmission structure has been proposed in [11]. In addition, a novel programmable weight phased array architecture has been developed [12]. The authors in [13] have considered the polygon construction in the complex plane to alter the transmission weight vector at symbol rate to achieve secure transmission.

Due to the analog array transmission structure, the above methods can only generate one beam with secure directional modulation. This contradicts the case of RIS-aided networks where secure transmissions of at least two beams, i.e., the beams aligned to the RIS and the target user, shall be simultaneously achieved. Moreover, the interplay between the two beams can cause high sidelobe energy, which is a waste of energy and can increase the security risk. Nevertheless, it is generally non-trivial to tackle the above problem due to the discrete optimization in antenna subset selection.

To enable secure directional modulation for RIS-aided communication networks, this chapter considers the hybrid beamforming structure and proposes a cross-entropy iterative method to lower the sidelobe at the base station. Here, the maximum sidelobe energy is used as the objective function and the Kullback-Leibler divergence is deployed to iteratively select good candidate antenna subsets. According to the simulation results, a reduction of 8 dB is achieved with respect to the maximum sidelobe energy. Besides, the symbol error rate at the eavesdropper maintains 0.75/0.875 for QPSK/8PSK even when signal-to-noise ratio increases. To the best of our knowledge, this is the first time for a secure directional modulation technique to be incorporated in RIS-aided communication networks.

Notations: We let a, $\mathbf{a}$, $\mathbf{A}$ represent the scalar, vector, and matrix respectively; $(\cdot)^{\mathrm{T}}$, $(\cdot)^{\mathrm{H}}$, and $(\cdot)^{-1}$ denote the transpose, conjugate transpose, and inverse of a matrix, respectively. $\mathcal{CN}(\boldsymbol{\mu}, \boldsymbol{\Sigma})$ represents circular symmetric complex Gaussian distribution with mean $\boldsymbol{\mu}$ and covariance $\boldsymbol{\Sigma}$.

6.2 System Model

We consider the RIS-aided communication networks where secure directional modulation is incorporated to achieve physical-layer security. In the considered system, there exists a BS as well as a RIS which intelligently reflects the signals transmitted from gNB, as shown in Fig. 6.1. The gNB is assumed to use hybrid beamforming architecture with N_{RF} RF chains and N antennas, and the RIS is assumed to have M passive reflective elements. The distance between the elements in the antenna array

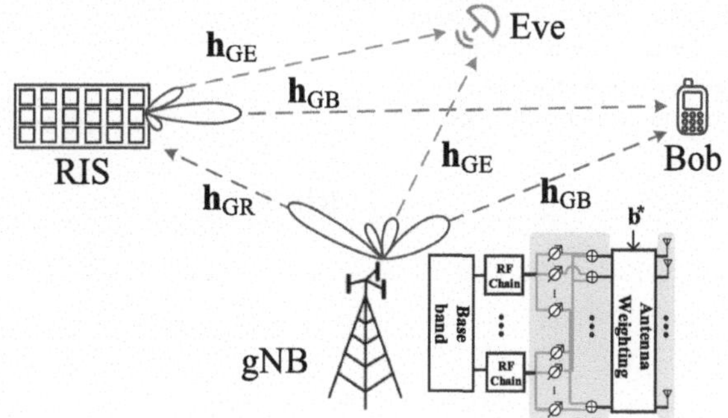

Fig. 6.1 Secure directional modulation for RIS-aided networks

is defined as d, with $d \le \lambda/2$ and λ is the wavelength. We assume that there are two users in the system. The legitimate user, indexed as Bob, is served simultaneously by gNB and RIS. However, the eavesdropper, indexed as Eve, may intercept the wireless signals emitted by gNB and RIS from any random unknown directions. Hence, secure directional modulation is incorporated in the considered system to achieve secure transmissions.

Denote the channel coefficients between gNB and the RIS, gNB and Bob, gNB and Eve, RIS and Bob, and RIS and Eve as $\mathbf{h}_{\mathrm{GR}} \in \mathbb{C}^{N \times M}$, $\mathbf{h}_{\mathrm{GB}} \in \mathbb{C}^{1 \times N}$, $\mathbf{h}_{\mathrm{GE}} \in \mathbb{C}^{1 \times N}$, $\mathbf{h}_{\mathrm{RB}} \in \mathbb{C}^{1 \times M}$, and $\mathbf{h}_{\mathrm{RE}} \in \mathbb{C}^{1 \times M}$, respectively. The phase shift matrix of the RIS is denoted as $\Phi = \mathrm{diag}\{\phi_1, ..., \phi_m, ..., \phi_M\} \in \mathbb{C}^{M \times M}$, where $\phi_m = e^{j\theta_m}$ and $\theta_m \in [0, 2\pi]$ is the m-th phase shift element of RIS. The received signal at Bob and Eve can be modeled as

$$y_{\mathrm{B}} = (\mathbf{h}_{\mathrm{RB}}\Phi\mathbf{h}_{\mathrm{GR}} + \mathbf{h}_{\mathrm{GB}})\mathbf{x}_{\mathrm{s}} + n_{\mathrm{B}}, \tag{6.1}$$

and

$$y_{\mathrm{E}} = (\mathbf{h}_{\mathrm{RE}}\Phi\mathbf{h}_{\mathrm{GR}} + \mathbf{h}_{\mathrm{GE}})\mathbf{x}_{\mathrm{s}} + n_{\mathrm{E}}, \tag{6.2}$$

where $\mathbf{x}_{\mathrm{s}} = \mathbf{F}_{\mathrm{RF}}\mathbf{F}_{\mathrm{BB}}x \in \mathbb{C}^{N \times 1}$ is the transmit signal on the antenna array of gNB, $\mathbf{F}_{\mathrm{RF}} \in \mathbb{C}^{N \times N_{\mathrm{RF}}}$ and $\mathbf{F}_{\mathrm{BB}} \in \mathbb{C}^{N_{\mathrm{RF}} \times 1}$ are respectively the analog and digital beamforming vectors, and $x = \sqrt{E_s}e^{j\varphi_k}$ is the source modulation symbol with $\sqrt{E_s}$ the signal amplitude and φ_k the phase shift. We further assume that $\|\mathbf{F}_{\mathrm{RF}}\mathbf{F}_{\mathrm{BB}}\|^2 = \rho$ where ρ is the total transmit power. The additive white Gaussian noises are with zero mean and $N_0/2$ variance, i.e., $n_{\mathrm{B}}, n_{\mathrm{E}} \sim \mathcal{CN}(0, \frac{N_0}{2})$.

6.3 Proposed CE-Based Hybrid Beamforming Scheme for Secure Transmissions

In this section, we aim to design secure directional modulation scheme for RIS-aided communication networks under the hybrid beamforming structure. To achieve this, gNB should firstly generate two secure steering beams targeting the directions of the

Fig. 6.2 Schematic diagram of signal transmission field strength for RIS-aided communication networks

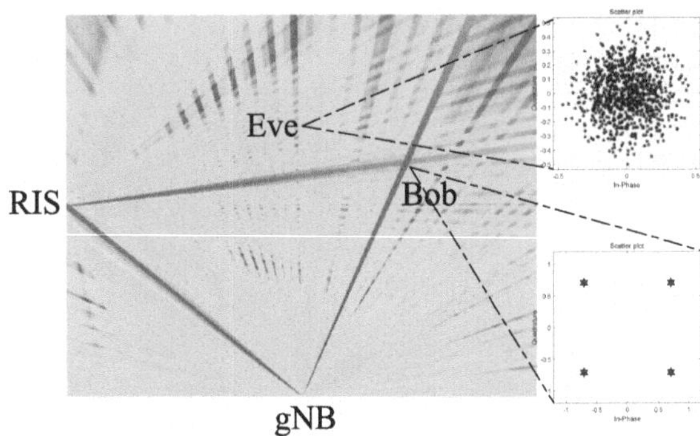

legitimate user Bob and the RIS, such that the predefined constellation diagram is transmitted within the two directions while the potential eavesdropper at any other direction only receives randomized signals. Upon receiving the predefined constellation from gNB, the RIS reflects the signals to Bob in another steering beam also using secure directional modulation, as illustrated in Fig. 6.2.

The challenges raised in the above scheme are two-folds: how to generate the secure steering beams, as well as how to mitigate the high sidelobe energy emission caused by the interplay between these beams while maintaining physical-layer security. The following parts propose an efficient two-stage hybrid beamforming algorithm to fulfill the above goals. The first stage focuses on the generation of a single secure beam via analog precoding. The second stage jointly considers the two secure beams and incorporates a cross-entropy method to mitigate the sidelobe under the the hybrid precoding structure.

6.3.1 Stage-1: Secure Steering Beam Generation

Denote the steering vector of gNB towards a receiver located in the direction of θ as

$$\mathbf{h}(\theta)^H = [e^{-j(\frac{N-1}{2})\frac{2\pi d}{\lambda}\cos\theta}, ..., e^{j(\frac{N-1}{2})\frac{2\pi d}{\lambda}\cos\theta}]. \tag{6.3}$$

Define $\mathbf{w} \in \mathbb{C}^{1 \times N}$ as the analog beamforming vector towards the direction θ. Under the principle of secure directional modulation, $\mathbf{w}$ can be derived as

$$\mathbf{w} = \frac{1}{S}[\mathbf{b} \odot \mathbf{h}(\theta)], \tag{6.4}$$

where $\mathbf{b} = [b_1, \ldots, b_n, \ldots, b_N]^T \in \{1, k\}^N$ is the pseudo-random weight vector over N antennas, $k \in \mathbb{R}$ is the weight coefficient, $\odot$ denotes the element-wise multiplication, and $S > 0$ is the normalizing factor. We assume that there are

$N - K$ antennas holding the weight of k and others 1, and S is given by $S = M + (N - K)|k|$.

For each time slot, the received signal at direction θ_T is

$$y = \mathbf{h}(\theta_T)\,\mathbf{w}x + n = \rho(\theta, \mathbf{b})x + n, \tag{6.5}$$

where $x \in \mathbb{C}$ and $n \in \mathbb{C}$ are the source symbol and the noise, respectively. The term $\rho(\theta, \mathbf{b})$ is defined as the scaling factor over x, where $\mathbf{b}$ changes for each slot. An approximate stochastic model of $\rho(\theta, \mathbf{b})$ can be derived as

$$\rho(\theta, \mathbf{b}) = \frac{1}{S} \sum_{n=0}^{N-1} b_n e^{j\left(n - \frac{N-1}{2}\right)\left(\gamma_\theta - \gamma_{\theta_T}\right)}, \quad \gamma_\theta = \frac{2\pi d}{\lambda} \cos\theta. \tag{6.6}$$

where adjusting $\mathbf{b}$ results in different beam characteristics.

For the conventional analog beamforming scheme, all antenna elements experience the same weight at any discrete-time, i.e., $\mathbf{b} = \mathbf{1}$ and $S = N$. Then $\rho(\theta, \mathbf{b})$ is re-written as

$$\begin{aligned}
\rho(\theta, \mathbf{b}) &= \frac{1}{N} \sum_{n=0}^{N-1} b_n e^{j\left(n - \frac{N-1}{2}\right)\left(\gamma_\theta - \gamma_{\theta_T}\right)} \\
&= \frac{1}{N} \frac{\sin\left(\frac{N\left(\gamma_\theta - \gamma_{\theta_T}\right)}{2}\right)}{\sin\left(\frac{\left(\gamma_\theta - \gamma_{\theta_T}\right)}{2}\right)}.
\end{aligned} \tag{6.7}$$

In this case, $\rho(\theta, \mathbf{b})$ is in the form of a real Sinc function, which causes the fact that the constellation received by Eve is exactly the same as Bob, with only beam gain degradation. If Eve deploys the large-aperture antenna or ultra-low sensitivity receiving technology, it is possible for Eve to recover the source symbols. This turns out to become a security threat in conventional analog beamforming.

For the secure directional modulation scheme, we control the weight of the transmitted signal through a programmable amplifier on each antenna element. Specifically, we deploy the optimized weight antenna subset transmission technique [12], which chooses $b_n = 1$ with probability p and $b_n = k$ with probability $1 - p$, such that b_n obeys Bernoulli distribution

$$b_n \sim \mathrm{Bern}(p) = \begin{cases} 1, & \text{probability}(p), \\ k, & \text{probability}(1 - p). \end{cases} \tag{6.8}$$

For large enough N, $\rho(\theta, \mathbf{b})$ obeys Gaussian distribution regardless of p, with mean $\mu(\theta)$ and variance $\sigma^2(\theta)$ derived as

$$\mu(\theta) = \frac{1}{N} \frac{\sin\left(\frac{N\left(\gamma_\theta - \gamma_{\theta_T}\right)}{2}\right)}{\sin\left(\frac{\left(\gamma_\theta - \gamma_{\theta_T}\right)}{2}\right)}, \tag{6.9}$$

and

$$\sigma^2(\theta) = \frac{(k-1)^2 p(1-p)}{N(p+k(1-p))^2},\qquad(6.10)$$

respectively. Based on the antenna weighting scheme, the eavesdropper in any undesired direction $\theta \neq \theta_T$ only receives the artificial noise, which guarantees the physical-layer security for all but the desired angle. Setting $k = \frac{1-K}{N-K}$ [12] can maximize the noise power and optimizes secure performance.

6.3.2 Stage-2: Sidelobe Reduction via Hybrid Beamforming

The following part jointly designs the digital and analog beamformers so as to reduce the sidelobe emission caused by the superposition of multiple analog beams.

In order to focus on both Bob and RIS, we set $N_{\mathrm{RF}} = 2$ and respectively generate two beams with two analog beamformers, i.e., $\mathbf{b} \odot \mathbf{h}_{\mathrm{GR}}^H$ and $\mathbf{b} \odot \mathbf{h}_{\mathrm{GB}}^H$, according to (6.4) and (6.8). The overall analog beamformer at gNB is thus denoted as

$$\mathbf{F}_{\mathrm{RF}} = [\mathbf{b} \odot \mathbf{h}_{\mathrm{GR}}^H, \mathbf{b} \odot \mathbf{h}_{\mathrm{GB}}^H] = \mathbf{b} \odot [\mathbf{h}_{\mathrm{GR}}^H, \mathbf{h}_{\mathrm{GB}}^H].\qquad(6.11)$$

As for the reflection path, the secure analog beamforming is also adopted by RIS, and the overall phase shift matrix of RIS is denoted as

$$\Phi = \mathbf{b}_{\mathrm{R}} \odot \Phi_{\mathrm{R}},\qquad(6.12)$$

where $\mathbf{b}_{\mathrm{R}} \in \{1, -1\}^M$ is the weight vector over M passive elements, and $\Phi_{\mathrm{R}} = \mathrm{diag}\{\phi_1, ..., \phi_m, ..., \phi_M\}$ is the steering vector towards Bob.

Combining (6.11) and (6.12), the equivalent channel response between gNB and Bob, termly $\mathbf{h}_{\mathrm{eq}}$, can be written as follows

$$\mathbf{h}_{\mathrm{eq}} = (\mathbf{h}_{\mathrm{RB}} \Phi \mathbf{h}_{\mathrm{GR}} + \mathbf{h}_{\mathrm{GB}}) \mathbf{F}_{\mathrm{RF}}.\qquad(6.13)$$

On this basis, we deploy the maximum ratio transmit (MRT) beamforming for digital precoding where $\mathbf{F}_{\mathrm{BB}}$ is derived as

$$\mathbf{F}_{\mathrm{BB}} = \frac{\mathbf{h}_{\mathrm{eq}}^H}{\|\mathbf{h}_{\mathrm{eq}}\|}.\qquad(6.14)$$

We then consider the design of $\mathbf{b}$ in details. Although randomly generating $\mathbf{b}$ with Bernoulli distribution promises security, the unexpected interplay between the two beams aligned to RIS and Bob may cause high sidelobe. Observing that the radiation patterns are affected by the weight vectors, we develop a cross-entropy method to down select a candidate set of $\mathbf{b}$ within $\{1, k\}^N$ which holds lower sidelobe.

Algorithm 6.1 Hybrid Beamforming Scheme Based on Cross-Entropy Iterative Algorithm

---

Input: N_c, N_s, N_{iter}.

Initialization: Probability of each antenna to choose through: $\mathbf{p}_b^0 = 0.5 * \mathbf{1}_{1 \times N}$, Number of iterations: $i = 0$

1: **for** $0 \leq i \leq N_{\text{iter}} - 1$ **do**

2:    Randomly generate N_c antenna weight vectors $\{\mathbf{b}^n\}_{n=1}^{N_c} = \{\mathbf{b}^1, \mathbf{b}^2, \ldots, \mathbf{b}^{N_c}\}$ with probability $\mathbf{p}_b^i$;

3:    Calculate the analog beamforming vector of the N_c group antenna combination $\{\mathbf{F}_{\text{RF}}\}_{n=1}^{N_c}$ via (6.11);

4:    Calculate the equivalent channel $\{\mathbf{h}_{\text{eq}}\}_{n=1}^{N_c}$ according to $\{\mathbf{F}_{\text{RF}}\}_{n=1}^{N_c}$ via (6.13);

5:    Use MRT algorithm to calculate the digital beamforming vector $\{\mathbf{F}_{\text{BB}}\}_{n=1}^{N_c}$ via (6.14);

6:    Calculate the *Fitness* of N_c antenna weight vectors as $\eta_1^i, \eta_2^i, \ldots, \eta_{N_c}^i$ via (6.15);

7:    Sort the *Fitness* as $\eta_{\pi(1)}^i \leq \eta_{\pi(2)}^i \cdots \leq \eta_{\pi(N_c)}^i < \text{T}_{\text{H}}$ with $\pi(.)$ the sorting function;

8:    Update the pass-through probability with first N_s weight vectors: $\mathbf{p}_b^{i+1} = \frac{1}{N_s} \left(\mathbf{b}^{\pi(1)} + \ldots + \mathbf{b}^{\pi(N_s)} \right)$;

9:    $i = i + 1$.

10: **end for**

Output: $\mathbf{b}^*$.

First, define a *Fitness* function as the optimization objective

$$\textit{Fitness} = \max(|\rho(\theta_1, \mathbf{b})|, \ldots, |\rho(\theta_n, \mathbf{b})|, \ldots, |\rho(\theta_N, \mathbf{b})|), \tag{6.15}$$

where $\rho(\theta_n, \mathbf{b})$ denotes the equivalent channel gain at direction θ_n given $\mathbf{b}$. It follows that (6.15) illustrates the maximum value of the sidelobe emissions in all undesired directions.

Then the sidelobe reduction problem is formulated as choosing $\mathbf{b}^*$ that holds the low sidelobe emission

$$\mathbf{b}^* = \underset{\mathbf{b} \in \{1, k\}^N}{\arg\min} (\textit{Fitness} < \text{T}_{\text{H}}), \tag{6.16}$$

where $\text{T}_{\text{H}} > 0$ is a predefined threshold, such that the sidelobe generated by $\mathbf{b}^*$ is not higher than T_{H} even when Eve's direction is not determined.

While solving the above discrete optimization problem is non-trivial, we then resort to the iterative random sampling approach and consider a cross-entropy method. The considered method generally iterates between two stages: (1) Generate random candidate solutions based on a parameterized probability distribution. (2) Evaluate the candidate solutions using the *Fitness* function and update the parameters of the probability distribution such that satisfactory solutions can be generated during the next iteration.

The details of the proposed cross-entropy method are provided in Algorithm 1. For initialization, the pass-through probability vector is initialized to be $\mathbf{p}_b^0 = 0.5 * \mathbf{1}_{1 \times N}$, which indicates the probabilities of the antennas being selected. The sizes of the candidate solution set and the selected set are given as N_c and N_s, respectively, with $N_c > N_s$. Next, we randomly generate N_c antenna weight vectors according to the pass-through probability. The *Fitness* values generated by N_c candidate $\mathbf{b}$s are then obtained based on (6.12) to (6.15), denoted as $\eta_1^i, \eta_2^i, \ldots, \eta_{N_c}^i$, respectively. After sorting the *Fitness* values, we take N_s largest $\mathbf{b}$s and update the probability $\mathbf{p}_b^{i+1}$ for the next iteration. Until the end of iteration, we can output $\mathbf{b}$ with optimized probability.

While the uniqueness of the above problem is not required, we define Ω as the collection of all generated solutions $\mathbf{b}^*$. During on-site deployment, the weight vector is randomly extracted out of Ω and applied at the antenna array to simultaneously realize security and low sidelobe emission.

The algorithm proposed above can also be applied at RIS to design the analog beamforming vectors, except that the weight of the antenna is $b_n = 1$ or -1.

6.4 Simulation Results

This section shows the sidelobe emission at gNB as well as the security performance of the proposed RIS-aided secure directional modulation method. The overall system setting is illustrated in Fig. 6.2 and the parameters used in the simulations are listed in Table 6.1.

Figure 6.3 compares the radiation patterns obtained by the hybrid beamforming with the proposed cross-entropy method and the conventional random antenna subset selection method [14]. It is shown that, the maximum sidelobe energy of the proposed scheme is approximately -23 dB. As a comparison, a 8 dB gain is achieved compared to the random selection method, which as a result reduces the energy received by Eve.

Figure 6.4 compares the empirical probability distribution functions of the sidelobe energy. The radiation pattern based on the proposed scheme and traditional scheme are calculated and normalized by subtracting the mainlobe energy. Then the

Table 6.1 Simulation parameters

Parameter	Symbol	Value
No. of antennas at gNB	N	128
No. of RF chains	N_{RF}	2
No. of passive elements at RIS	M	64
No. of candidate solutions	N_c	100
No. of selected solutions	N_s	25
No. of iterations	N_{iter}	70
Angle of Bob (degree)	θ_{Bob}	110
Angle of RIS (degree)	θ_{RIS}	45
Angle of Eve (degree)	θ_{Eve}	100

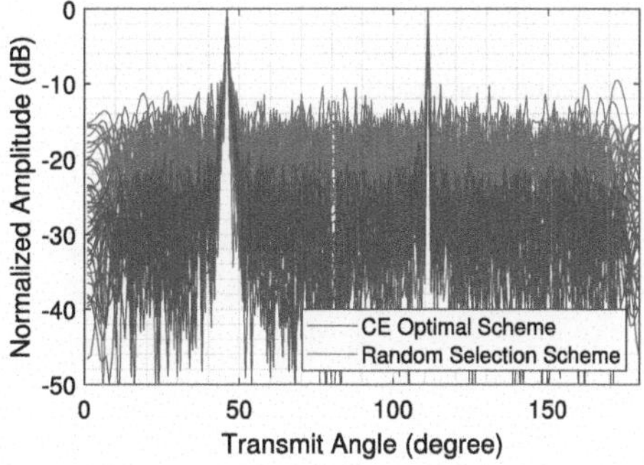

Fig. 6.3 Comparisons on the sibelobe emissions

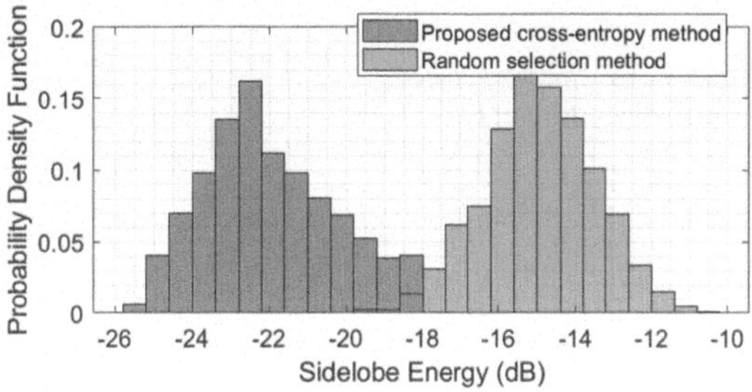

Fig. 6.4 Comparisons on the sidelobe energy emission

probability density distribution function of the maximum sidelobe energy is obtained through Monte Carlo simulations. It can be seen that the sidelobe energy obtained by the traditional scheme is relatively large, mainly distributed around -15 dB, while the sidelobe energy of the proposed scheme is 8 dB smaller than the traditional scheme.

Figure 6.5 shows the symbol error rate performance at Eve. The transmission without secure directional modulation, termly pure QPSK/8PSK, can achieve low error rate at high SNR. This indicates that with high-cost receiver implementation, e.g. massive antennas, at Eve, security is not ensured. As a comparison, we set

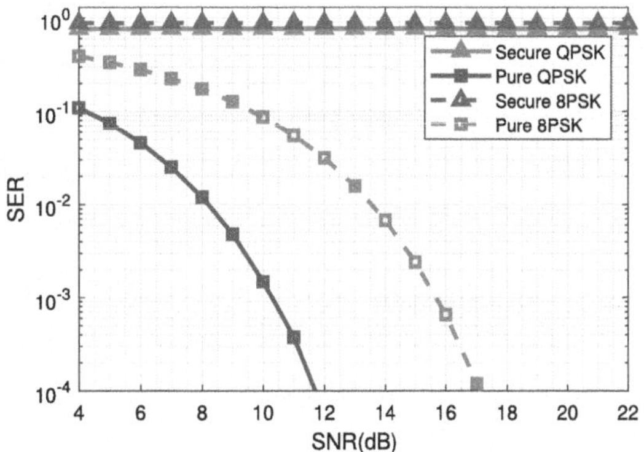

Fig. 6.5 SER performance at Eve. Secure/pure QPSK: with/without proposed secure transmission method

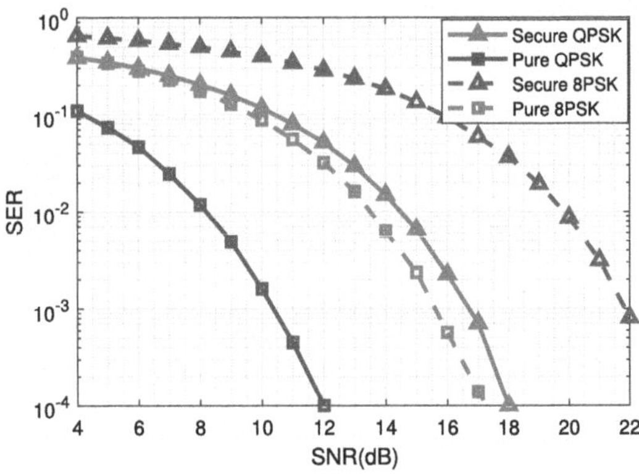

Fig. 6.6 SER performance at Bob. Secure/pure QPSK: with/without proposed secure transmission method

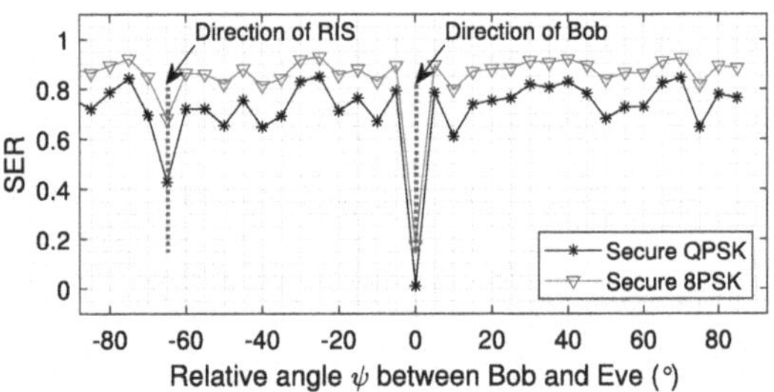

Fig. 6.7 SER performance of Eve under different ψ

$K = 64$ and $k = -1$ for the proposed secure directional modulation scheme, termly secure QPSK/8PSK, and show the error rate performance. The constellation diagram received by Eve is the same as the noise, as illustrated in Fig. 6.2. The symbol error rates of QPSK and 8PSK transmissions maintain at about 0.75 and 0.875, respectively, even if SNR increases. Figure 6.6 shows the relationship between the symbol error rate and SNR at the target user Bob. As SNR at Bob increases, the symbol error rate decreases. Since a portion of the antennas is selected to transmit with $k = -1$ in the proposed scheme, the total energy of the signal in the desired direction is reduced, so that the symbol error rate performance slightly degrades compared to pure QPSK/8PSK. In Fig. 6.7, we show the SER of Eve when the relative angle ψ between Bob and Eve changes arbitrarily. The results indicate that, except when lying exactly at the direction of Bob or RIS, the SER of Eve maintains a high level even if its angle changes arbitrarily. This validates the advantage of the proposed scheme that the security is achieved in any undesired directions.

6.5 Conclusions

In this chapter, the secure directional modulation based on hybrid beamforming was studied for RIS-aided communication networks. A cross-entropy iterative algorithm was proposed to optimize the hybrid beamformers at gNB and the analog beamformer at the RIS, so as to achieve physical-layer security while reduce the sidelobe emission. Simulation results validated that the proposed secure transmission scheme effectively reduced the sidelobe energy for more than 8 dB. Besides, for QPSK/8PSK transmissions, the symbol error rate at Eve in any undesired directions was approximately 0.75/0.875, which can meet the requirement of total security.

References

1. Q. Wu, R. Zhang, Towards smart and reconfigurable environment: intelligent reflecting surface aided wireless network. IEEE Commun. Mag. **58**(1), 106–112 (2020)
2. I. Trigui, W. Ajib, W.-P. Zhu, Secrecy outage probability and average rate of RIS-aided communications using quantized phases. IEEE Commun. Lett. **25**(6), 1820–1824 (2021)
3. M.H. Khoshafa, T. Ngatched, M.H. Ahmed, Reconfigurable intelligent surfaces-aided physical layer security enhancement in D2D underlay communications. IEEE Commun. Lett. **25**(5), 1443–1447 (2021)
4. Z. Tang, T. Hou, Y. Liu, J. Zhang, C. Zhong, A novel design of RIS for enhancing the physical layer security for RIS-aided NOMA networks. IEEE Wirel. Commun. Lett., 1 (2021)
5. L. Yang, J. Yang, W. Xie, M.O. Hasna, T. Tsiftsis, M.D. Renzo, Secrecy performance analysis of RIS-aided wireless communication systems. IEEE Trans. Veh. Technol. **69**(10), 12296–12300 (2020)
6. Z. Zhang, C. Zhang, C. Jiang, F. Jia, J. Ge, F. Gong, Improving physical layer security for reconfigurable intelligent surface aided NOMA 6G networks. IEEE Trans. Veh. Technol. **70**(5), 4451–4463 (2021)
7. J. Liu, J. Zhang, Q. Zhang, J. Wang, X. Sun, Secrecy rate analysis for reconfigurable intelligent surface-assisted MIMO communications with statistical CSI. China Commun. **18**(3), 52–62 (2021)
8. N.K. Kundu, M.R. McKay, RIS-assisted MISO communication: optimal beamformers and performance analysis, in *2020 IEEE Globecom Workshops (GC Wkshps)*, pp. 1–6 (2020)
9. I. Trigui, W. Ajib, W.P. Zhu, Secrecy outage probability and average rate of RIS-aided communications using quantized phases. IEEE Commun. Lett. **PP**(99), 1 (2021)
10. N. Valliappan, A. Lozano, R.W. Heath, Antenna subset modulation for secure millimeter-wave wireless communication. IEEE Trans. Commun. **61**(8), 3231–3245 (2013)
11. N.N. Alotaibi, K.A. Hamdi, Switched phased-array transmission architecture for secure millimeter-wave wireless communication. IEEE Trans. Commun. **64**(3), 1303–1312 (2016)
12. Y. Hong, X. Jing, H. Gao, Programmable weight phased-array transmission for secure millimeter-wave wireless communications. IEEE J. Sel. Topics Signal Process. **12**(2), 399–413 (2018)
13. X. Zhang, X.-G. Xia, Z. He, X. Zhang, Phased-array transmission for secure mmwave wireless communication via polygon construction. IEEE Trans. Signal Process. **68**, 327–342 (2020)
14. N. Valliappan, A. Lozano, R.W. Heath, Antenna subset modulation for secure millimeter-wave wireless communication. IEEE Trans. Commun. **61**(8), 3231–3245 (2013)

Chapter 7
OFDM Radar Range Accuracy Enhancement

In this chapter, we explore the integration technology of communication and sensing. Section 7.1 gives the background and motivation of OFDM radar design. Section 7.2 introduces the OFDM radar operation principle and the enhancement techniques. Section 7.3 shows the implementation detail of the proposed OFDM radar. Section 7.4 discusses the Interference from other radar. Section 7.5 presents a laboratory-based experimental setup and the measured result. Section 7.6 summarizes this chapter.

7.1 Introduction

There are two hot topics within information technology nowadays: one is Industry 4.0, and the other is 5G. Industry 4.0 was first proposed in 2011, focusing on connecting all machines with robotics in the factory to form an agile, reconfigurable factory which can be used to produce diverse commodities on demand [1]. 5G mobile infrastructure also promises high data rate, low latency and a higher quality of service [2, 3], which will undoubtedly will accelerate the development of the future smart factory by providing massive low latency wireless connections which can bring factory monitoring and control to a new level. Moveover, sensors play a crucial role in realizing fully configurable robotic tools in the future factory. In addition to robotic motion monitoring, it is also important to monitor human workers to avoid robotic arm and worker collision. A 5G C-band radar was demonstrated for human motion detection in [4, 5]. High-precision sensors also play a decisive role in the production of sophisticated equipment such as robotic arms, computer numerical control (CNC) lathes, milling machines, etc.

Past decades have witnessed the application of various sensors, such as accelerometers, temperature/humidity meters, lasers and ultrasound sensors in factories. References [6, 7]. Robotic arms will definitely play prominent roles in advanced automation [8]. One critical issue is to prevent these robotic arms from colliding with each

J. Li et al., *Key Technologies of High Frequency Wireless Communications*,
https://doi.org/10.1007/978-981-96-5894-7_7

Table 7.1 Demographic prediction performance comparison by three evaluation metrics

References	Freq. (GHz)	Bandwidth	Method	Accuracy (μm)	Unambiguous range	Measurement rate
[10]	80	20 GHz	FMCW	±8	≥5 m	0.25 kHz
[11]	121	6G Hz	FMCW	±6	≥1 km	62.77 Hz
[12]	122.5	1 GHz	FMCW	±2	≥1 km	1 kHz
[13]	80	10 GHz	FMCW	±0.5	≥1 km	250 Hz
[14]	24	50 kHz	Interferometer	±0.5	≥1.25 cm	≤10 kHz
[15]	24	250 MHz	2-tone Interferometer	±35	≥50 cm	–
This	79	2 GHz	OFDM and FRFT	±20	≥1 km	5 MHz

other, as well as to ensure that a robotic arm can manipulate objects with ultra-high accuracy. To measure the distance between robotic arms and that between the arm and the object under manufacture, several sensors are used including laser, ultrasonic and radar [9]. Laser requires a clear view of the target, which is not viable in some applications. While ultrasonic sensors are of relatively low cost, it is hard to avoid interference when multiple sensors are installed close to each other. A radar system with high distance accuracy and interference resilience is of great interest for high-precision robotic arm manipulation and to avoid collision between robotic arms.

Various radar systems were adopted for high-precision applications with their performance summarized in Table 7.1. In the literature [10], an unambiguous phase-based frequency modulated continuous wave (FMCW) radar algorithm for micron accuracy distance measurements was proposed. The unambiguous single measurement accuracy has been improved from about 400 μm to 8 μm at 80 GHz. The authors in [11] proposed a highly miniaturized and commercially available millimeter wave radar sensor working in the 121–127 GHz range, which can be utilized for micrometer range accuracy. FMCW radars at 120 GHz band have achieved micrometer level range accuracy [12, 13], but the measurement rate is often limited by the frequency sweep speed. Interferometer techniques with single tone or dual tone can be used to measure targets at high rate, but range ambiguity appears when the target moves out of a limited range window [14, 15]. Due to the high frequency and short wavelength of the millimeter wave, ultra-high ranging accuracy can be achieved by determining the carrier phase, and OFDM can resist interference by means of code division. Therefore, the millimeter wave OFDM radar sensor achieves high-precision measurement under strong interference.

Orthogonal Frequency Division Multiplexing radar waveform has been proposed in [16], which can perform radar detection as well as communication simultaneously. Based on the application of diverse signature codes to each sensor, multiple radars can operate simultaneously free of interference. When applying fast Fourier transform (FFT), the range accuracy and resolution of OFDM radar are inversely proportional to the bandwidth B of the transmitted signal, as $\delta R = c_0/2B$, where c_0 is the speed of light.

OFDM radar, with its multi-carrier parallel transmission and reception characteristics, has attracted wide attention in high-precision ranges. Random OFDM pair
radar, based on compressive sensing with high resolution range reconstruction, was
proposed in [17, 18]. Meanwhile, adaptive OFDM radar waveform design methods
were proposed for improved micro-Doppler estimation in [19, 20]. Then, the authors
in [21] proposed an OFDM integrated radar and a communication waveform design
method to improve the effectiveness of limited spectral resources. In addition, a two-
step Doppler estimation scheme was introduced in [22]. A novel signal processing
approach for OFDM radar has been presented in [23], which can overcome the effects
of Doppler in OFDM. Furthermore, a high accuracy Doppler processing method for
the OFDM radar and communication system has been introduced in [24].

In this chapter, we proposed an OFDM radar accuracy enhancement technique,
which provides an improvement in accuracy of 500-times compared to a traditional
FFT estimation result. With a bandwidth of B = 2 GHz, the distance measurement of
a metal target has an error of less than 20 μm. This result is experimentally verified
using an OFDM radar operating at 80 GHz.

7.2 Review of OFDM Radar Concept and Range Accuracy Limitation

The operational principle of an OFDM radar has been discussed in detail in [16].
An OFDM multicarrier signal can be denoted as a composite of multiple parallel
orthogonal signal carrier signals, modulated with different transmission data. The
transmission data can be written as:

$$x(t) = \sum_{\mu=0}^{N_{sym}-1} \sum_{n=0}^{N_c-1} a(\mu N_c + n)\exp(j2\pi f_n t)\mathrm{rect}(\frac{t - \mu T_{OFDM}}{T_{OFDM}}), \quad (7.1)$$

where μ is the OFDM symbol index within the total amount of N_{sym} OFDM symbols,
n is the subcarrier index within the total amount of N_c subcarriers, a is the complex
modulation symbol, f_n is the subcarrier frequency and T_{OFDM} is the OFDM symbol
duration.

In this paragraph, we introduce an intuitive explanation that is illustrated in
Fig. 7.1, complimenting the equations expressed in [16]. Assuming a transmission
waveform comprises N_{sym} symbols, each with a symbol duration of T_{sym}, then each
of the symbols is a sum of N_c subcarriers with a subcarrier spacing of $\Delta f = 1/T_{sym}$.
The initial phase of each subcarrier at each symbol is defined in the red matrix as
shown in Fig. 7.1a. Here we define a matrix:

$$T = \begin{bmatrix} t(1, 1) & \cdots & t(1, N_{sym}) \\ \vdots & \ddots & \vdots \\ t(N_c, 1) & \cdots & t(N_c, N_{sym}) \end{bmatrix}. \quad (7.2)$$

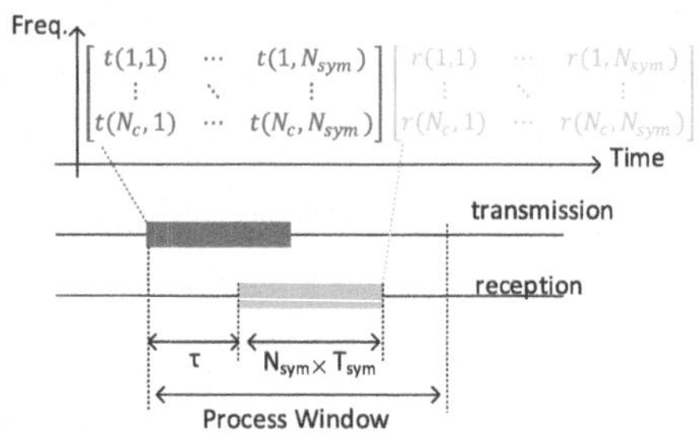

(a) Time domain waveform of transmitted and received radar signal.

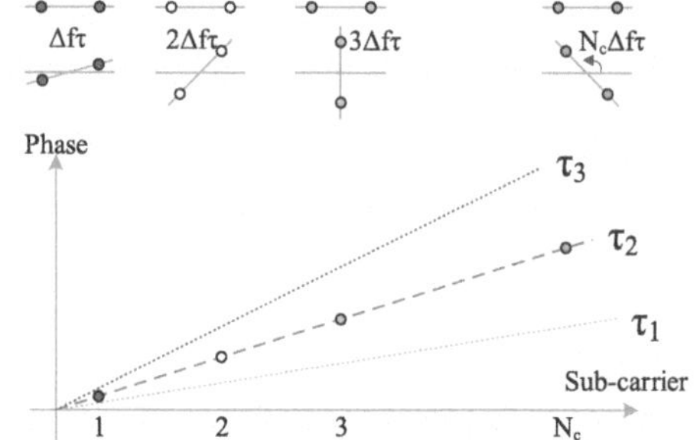

(b) Illustration of the phase rotation at each subcarrier.

Fig. 7.1 The intuitive explanation of OFDM radar

Each element in matrix T is a complex number to represent the phase of transmission: the first index denotes the subcarrier number and the second index denotes the symbol number. The red waveform in Fig. 7.1a contains $N_c \times N_{sym}$ symbols. For each transmitter, the matrix T is fixed as a signature code. In this work, BPSK modulation is used, therefore $t(m, n) \in [0, \pi]$. If the receiver is fully synchronized with the transmitter, the echo reflected from a target would generate a green waveform which is delayed by τ as shown in Fig. 7.1a. The method of estimating this delay with high accuracy will be focused in this chapter. With an observation window that is designed to be wide enough, the reception echo can be captured completely. Since the receiver is fully synchronized with the transmitter, the received signal matrix can be constructed similar to Eq. 7.2. The received signal matrix R can be written as:

$$R = \begin{bmatrix} r(1, 1) & \cdots & r(1, N_{sym}) \\ \vdots & \ddots & \vdots \\ r(N_c, 1) & \cdots & r(N_c, N_{sym}) \end{bmatrix}. \tag{7.3}$$

Note that the received signal contains the delay information that is of interest. By dividing matrix R with matrix T, the signal transmission channel can be modelled in both time and frequency domains using a matrix $D = R/T$. Such a matrix was described as Eq. 16 in [16] which is rewritten here as:

$$D = \begin{bmatrix} d(1,1) & \cdots & d(1,N_{sym}) \\ \vdots & \ddots & \vdots \\ d(N_c,1) & \cdots & d(N_c,N_{sym}) \end{bmatrix}. \tag{7.4}$$

The matrices T and R are illustrated in Fig. 7.1b for a more intuitive understanding. On the top of Fig. 7.1b, four BPSK constellation diagrams are shown with different colours, which correspond to subcarrier 1, subcarrier 2, subcarrier 3 and subcarrier N_c, respectively. For instance, the constellation diagram of subcarrier 1 corresponds to the first row of the matrix T. The constellation diagrams below are the received constellation for different subcarriers. Similarly, each constellation diagram corresponds to a certain row in matrix R. It can be noticed that when a single target is presented, the constellation at higher subcarriers is more rotated than those at lower subcarriers. The matrix D is the channel information as discussed before, each row in which represents the constellation rotation measurement at each subcarrier. Such measurements are made at the reception of every symbol, thus there are N_{sym} measurements in each row. Averaging can be made over each row thus matrix D converts into vector:

$$D' = \begin{bmatrix} \overline{d_1} \\ \overline{d_2} \\ \vdots \\ \overline{d_{N_c}} \end{bmatrix}, \tag{7.5}$$

where $\overline{d_m}$ is a complex number of the mean rotation of the constellation diagram at m-th subcarrier. Vector D' is plotted at the bottom of Fig. 7.1b with angles along the Y-axis and the frequencies of each subcarrier along the X-axis. As mentioned before with an ideal single target, the rotation angle increases in a linear manner as the subcarrier frequency increases, so the rotation angle at k-th subcarrier can be written as:

$$\varphi_k = 2\pi \times (f_c + k \times \Delta f) \times \tau. \tag{7.6}$$

A range estimation method is proposed in [16] using inverse discrete Fourier transform (IDFT) operations:

$$r(k) = IDFT[D'] = \frac{1}{N_c} \sum_{n=0}^{N_c-1} D'(n) exp(\frac{j2\pi}{N_c}nk)$$

$$= \frac{1}{N_c} \sum_{n=0}^{N_c-1} exp(-j2\pi n \Delta f \tau) exp(\frac{j2\pi}{N_c}nk). \tag{7.7}$$

It can be seen that two exponential terms would cancel each other and result in unity when:

$$k = \lfloor \Delta f N_c \tau \rfloor, k = 0, ..., N_c - 1. \tag{7.8}$$

This means a peak would occur at an index of k in the time response of $r(k)$ which indicates a target is detected at that position. The accuracy of using such a method is limited by the fact that IDFT is used. It can be seen that target range searching is made in a discrete step of $\delta R = \frac{C_0}{2N_c \Delta f}$.

This is illustrated in Fig. 7.1b as several dotted reference lines with different slopes are used to compare the observed rotation angles. Intuitively, using more reference lines with finer steps would give a more accurate estimation. A Fractional Fourier Transform based method is introduced in the following section to improve the estimation accuracy.

7.3 FRFT and Phase Analysis Techniques

As mentioned above the accuracy of the range resolution is a result of using a conventional Fourier transform as shown in Eq. 7.7. In this chapter we have proposed a two-step signal processing method that can improve range accuracy to micrometer level. The first-step is using fractional Fourier transformation (FRFT) instead of Fourier transform. The second-step uses the phase information of a carrier to further enhance estimation accuracy. Both steps are introduced and discussed in this section.

7.3.1 Range Accuracy Enhancement with FRFT

As mentioned in previous section, modifying the Eq. 7.7 by simply adding more searching steps will increase the estimation accuracy. By introducing a fractional factor ma, the estimation function using FRFT can be written as:

$$r(k) = \frac{1}{ma \times N_c} \sum_{n=0}^{ma \times N_c - 1} \exp(-j2\pi n \Delta f \tau)\exp(\frac{j2\pi nk}{ma \times N_c}). \qquad (7.9)$$

The fractional factor defines ma times points are searched, which results in a correlation curve as shown in Fig. 7.2. The figure shows the simulation result using the FRFT method where the simulation scenario is: 128 subcarriers with spacing of 15.625 MHz, $ma = 20$, carrier frequency 80 GHz and target distance set at 240 mm and 243.75 mm.

When using Eq. 7.7, the target at these two positions results in identical peaks, indicating that the estimation distance is 75 mm. Clearly, such a small distance change cannot be tracked by the conventional method. By using the FRFT method, the estimation curves are shown in Fig. 7.2a. The red curve corresponds to the target at a position of 240 mm and the blue curve corresponds to the target at 243.75 mm. The zoomed in figure is shown in Fig. 7.2b. It can be seen that due to the fine accuracy of the estimation curve the target movement can be tracked with high confidence. A

Fig. 7.2 Simulated FRFT correlation peak when the target is at 240 mm and 243.75 mm

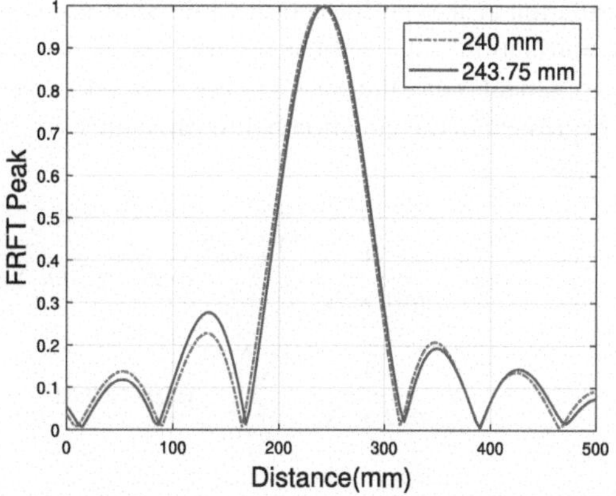

(a) Simulated peak of target at different locations.

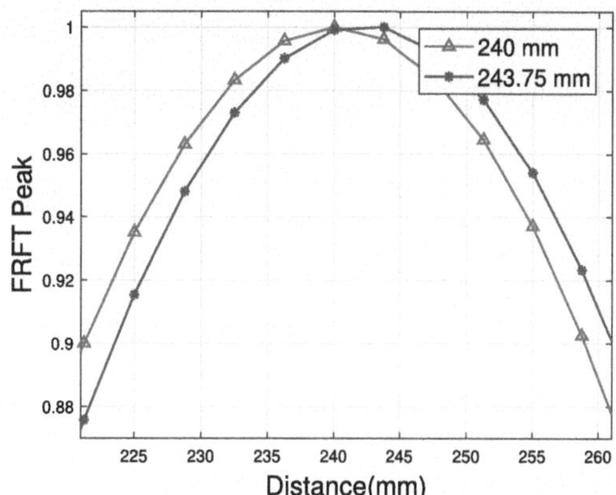

(b) Zoom-in view of the correlation peak.

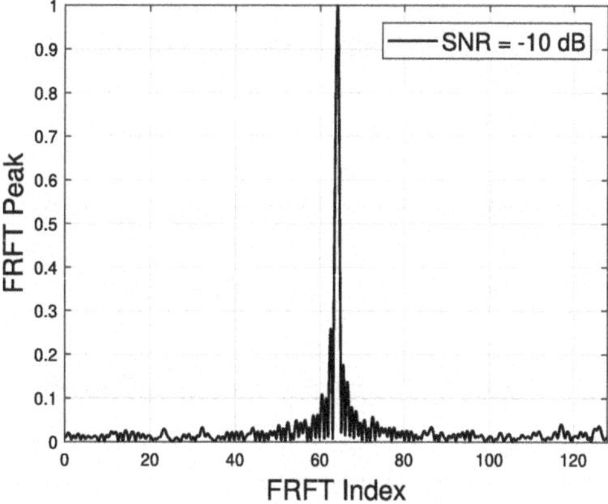

simulation is also performed in a low SNR scenario (SNR = -10 dB). As expected, the correlation peak is not as pronounced as it would be in the ideal case. To resolve this practical issue, the curves in Fig. 7.2c are correlated with an ideal "sinc" function as expressed in the equation below, then the correlation peaks appear again after this process:

$$
\begin{aligned}
r(k) &= \frac{1}{ma \times N_c} \sum_{n=0}^{ma \times N_c - 1} \exp(-j2\pi n \Delta f \tau)\exp(\frac{j2\pi nk}{ma \times N_c}) \\
&= \frac{1}{ma \times N_c} \sum_{n=0}^{ma \times N_c - 1} \exp(j2\pi n(\frac{k}{ma \times N_c} - \Delta f \tau)),
\end{aligned} \tag{7.10}
$$

where

$$
\Delta \tau = \frac{k}{ma \times N_c} - \Delta f \tau, \tag{7.11}
$$

Then the equation can be further simplified into

$$
\begin{aligned}
r(k) &= \frac{1}{ma \times N_c} \sum_{n=0}^{ma \times N_c - 1} \exp(j2\pi n \Delta \tau) \\
&= \frac{1}{ma \times N_c} \frac{1 - \exp(j2\pi \Delta \tau \times ma \times N_c)}{1 - \exp(j2\pi \Delta \tau)} \\
&= \frac{1}{ma \times N_c} \frac{\sin(\pi \Delta \tau \times ma \times N_c)}{\sin(\pi \Delta \tau)} \\
&\approx \frac{1}{ma \times N_c} \frac{\sin(\pi \Delta \tau \times ma \times N_c)}{\pi \Delta \tau} \\
&= \mathrm{sinc}(\Delta \tau \times ma \times N_c).
\end{aligned} \tag{7.12}
$$

The achievable range accuracy depends on several system parameters such as SNR, quadrature balance, in-band flatness, etc. In our system, the correlation method is only used for coarse estimation. As long as the estimation accuracy is better than the wavelength of the carrier, the phase estimation can be used for further improvement of the accuracy down to micrometer level. The phase-assisted estimation method is discussed in the following section.

7.3.2 Range Accuracy Enhancement with Phase Analysis

Taking advantage of the short wavelength of the millimeter wave carrier, phase rotation occurs even with micrometer target displacement. By examining the phase of the modulated carrier, the distance of the target can be tracked with micrometer level accuracy. Such a method is widely used by interferometer radars [14, 15]. However, ambiguity appears when the target moves more than one carrier wavelength.

In this chapter, as we demonstrated in the previous section, the proposed FRFT method can locate the target with an accuracy of less than half a wavelength. Adding phase analysis allows us to further improve the accuracy to micrometer level. The distance of the target can be written as follows:

$$R = m \times \lambda_c + R'. \tag{7.13}$$

For any subcarrier in the OFDM symbol, the target equals m-time carrier wavelength with a remaining distance R'. The range estimation $\widehat{R}_{OFDM}$ obtained by the FRFT method above is used for calculating the integer number m. The estimation error will be eliminated by a quantizing procedure. The remaining distance is calculated by:

$$R' = \frac{\phi}{2\pi} \times \lambda_c, \tag{7.14}$$

where ϕ is the received constellation rotation at the subcarrier of wavelength λ_c. At 80 GHz ($\lambda_c = 3.75$ mm), for example, a 0.1° rotation is equal to 1 μm target movement. The range estimation accuracy is then translated into angle resolution of the constellation rotation.

By using the proposed two-step method, the distance of the target can be estimated using Eq. 7.14 with a wide unambiguous window. Experimental validation of this method is presented in the following section.

7.4 Multi-radar Interference Cancellation

The proposed OFDM radar utilizes a signature code when sensing the target. With multiple signature codes, radars can be installed closely without interfering with each other. In this section, the principle and performance benchmarks are presented with an analysis of a scenario with multiple radars coexisting.

As mentioned above, the radar transmits a specific symbol sequence as T, shown in Eq. 7.2. By assigning orthogonal symbol sequences to different radars, the interference between radars can be minimized. In our work, the selected transmitting symbol sequence is a m-sequence generated by polynomial $p(x) = x^7 + x^6 + 1$ using linear feedback shift register techniques. By assigning different m-sequence codes to different radars, radars can work simultaneously. To understand the performance of radars under interference, simulations are conducted using the situation in which a radar is operated under the influence of another radar. The simulation result is presented in Fig. 7.3.

In this simulation, 128-bit m-sequences are assigned to two radars which operate at the same frequency. The target range error is plotted versus received SNR. First,

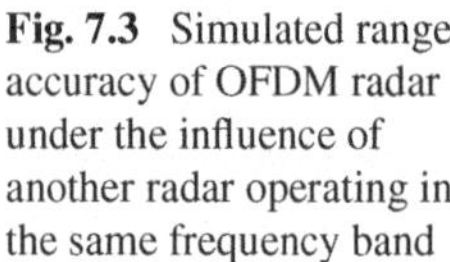

Fig. 7.3 Simulated range accuracy of OFDM radar under the influence of another radar operating in the same frequency band

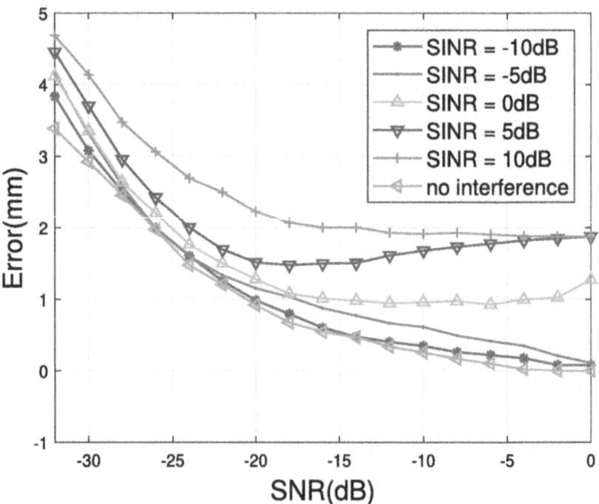

when only one radar is operated, the estimation error is plotted as "no interference". Then another in-band operated radar is added with the signal-to-interference-plus-noise ratio (SINR) varying from -10 dB to 10 dB. The simulation results are as follows:

- An estimation error of less than half of the carrier wavelength (79 GHz, 1.85 mm) can be achieved even at -22 dB SNR with the SINR of 5 dB.
- Due to the fact that estimation is made by averaging 128 sub-carriers, the proposed radar is relatively immune to in-band interference. When the SINR = 10 dB, the radar can also work well.
- The simulation shows that at SINR = 5 dB, estimation with half of the wavelength can still be achieved at SNR= -22 dB. This means that the range accuracy can be further improved by phase analysis at SNR = -22 dB when the SINR = 5 dB.

7.5   Experimental Result

The proposed OFDM radar is experimentally verified using commercially available transceiver modules at 79 GHz in a laboratory environment. In this section, the experimental setup is introduced and measurement results are presented and discussed.

7.5.1   Experimental Setup

The photo of the experimental laboratory setup is shown in Fig. 7.4. Integrated E-band transmitter and receiver modules are used to generate and receive signals at 79 GHz.

Fig. 7.4 Photo of the radar experiment setup in the lab

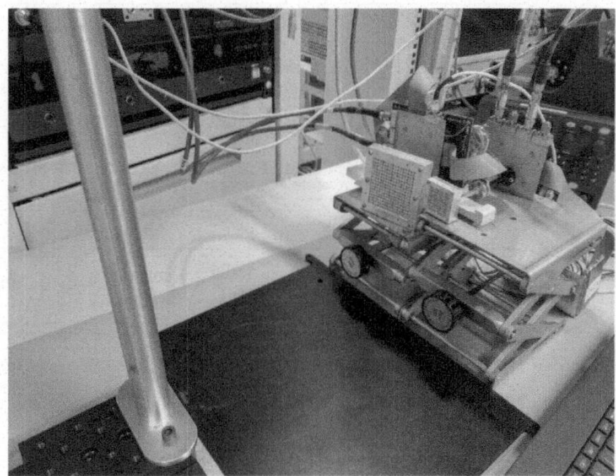

Fig. 7.5 Detailed diagram of the setup

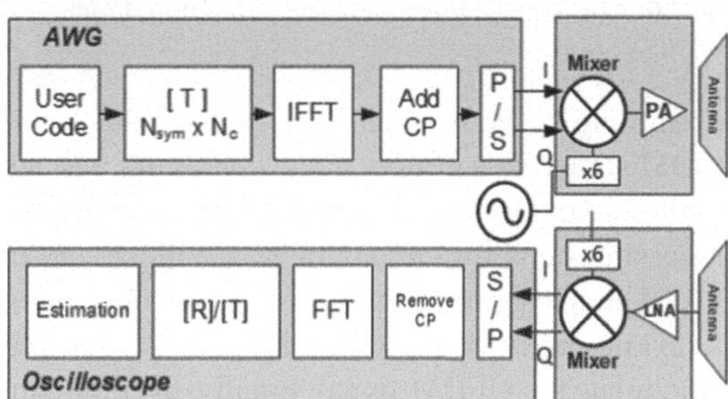

Two planar E-band antennas are used, with separate antennas, achieving more than 50 dB isolation between transceivers. A metal bar is used as a target which is located 24 cm away from the radar. An arbitrary waveform generator (AWG) is used to generate an OFDM signal and a real-time oscilloscope is used to capture the received signal. The received data is processed off-line using Matlab.

In Fig. 7.5, the detailed diagram of the setup is depicted. In the AWG, the user code is used to generate a transmission matrix T as in Eq. 7.2. 128 subcarriers are used with channel spacing of 15.625 MHz, therefore a total bandwidth of 2 GHz is occupied. The time domain waveform is generated using 1024 points inverse fast Fourier transform (IFFT), and the cyclic prefix (CP) is added. The waveform is then converted from parallel into serial (P/S) and output from AWG at 16 Gsps. An E-band transmitter (Gotmic gtsc0023) is used with 13 dBm average output power. A 30 dBi antenna is used on the transmitter side. A 25 dBi receiver antenna is used together with an E-band receiver (Gotmic gRsc0013). Transmitter and receiver share the same local oscillator (LO), therefore the phase between transmitter and receiver is fully synchronized. The received quadrature baseband signals are sampled by an oscilloscope.

The detailed transmitted pre-processing steps for OFDM radar are shown in Fig. 7.6a. The user code was generated at the start to distinguish each user. Then

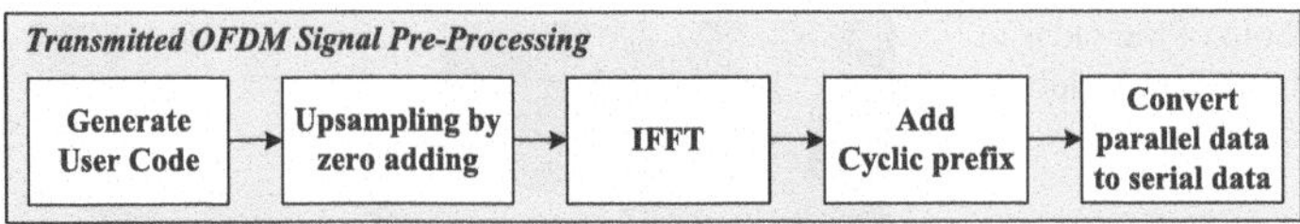

(a) The detailed transmitted pre-processing steps for OFDM radar.

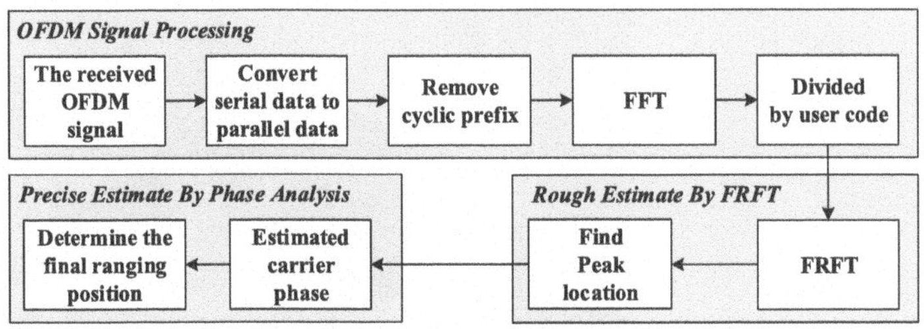

(b) The detailed received processing steps for OFDM radar.

Fig. 7.6 The detailed processing steps for OFDM radar

upsampling by zero adding was performed. Next, IFFT was used to generate the OFDM signal. In addition, the cyclic prefix was added to resist multipath problems. Finally, we converted the parallel data to serial data. Moreover, the detailed received processing steps for OFDM radar are illustrated in Fig. 7.6b. At the first stage, the OFDM signal should be processed. We convert serial data to parallel data, and then the cyclic prefix of the OFDM signal is removed. Then the FFT can be used to demodulate the OFDM signal. Finally, the phase information can be uniquely left by dividing by the user codes. At the second stage, the peak location of the FRFT can be used to perform a rough estimate as mentioned in section III. At the last stage, since the estimation accuracy of the second stage is already within one wavelength, the phase analysis can be adopted to further improve the range accuracy.

7.5.2 *Experimental Result*

The normalized received signal spectrum is shown in Fig. 7.7. The signal occupies 2 GHz bandwidth with a received SNR higher than 20 dB. The received power is -17 dBm when the target is 240 mm away. Considering the path loss, the radar cross-section (RCS) is -38.4 dB/mm^2.

The target is then moved from 240 mm to 248 mm with 1 mm steps using a manual micrometer positioner. The measured phase of different subcarriers is plotted in Fig. 7.8when target locates at 240 mm, 243 mm and 246 mm, respectively.

As described in section II, the phase slope is changed proportionally to the target distance as expected. Using the FRFT method, the target distance can be estimated as shown in Fig. 7.9. As long as the error of this estimation is less than the wavelength (3.7 mm), the distance can be further improved with phase analysis.

Fig. 7.7 The normalized
received signal spectrum

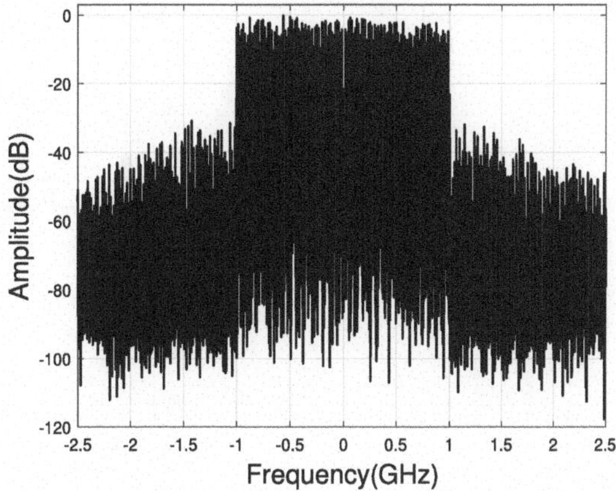

Fig. 7.8 The measured
phase of different subcarriers

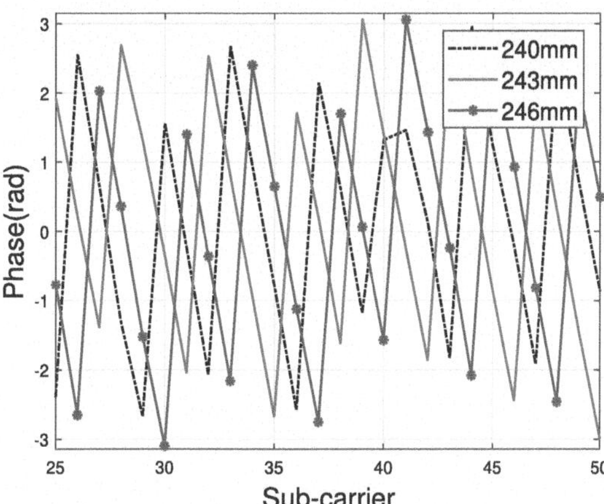

Fig. 7.9 The estimated
target distance by the FRFT
method

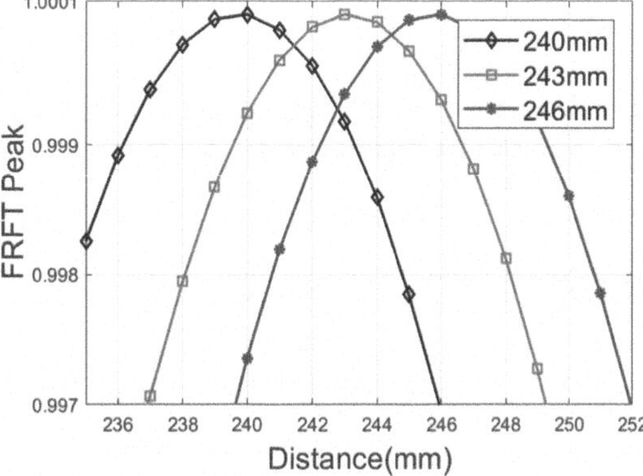

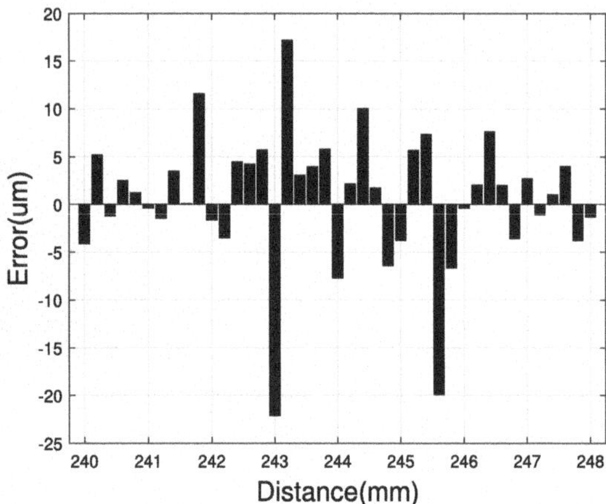

Fig. 7.10 The final estimated result by the phase analysis method

Adding phase analysis mentioned in section III. B, the final estimation result is presented in Fig. 7.10. It can be seen that the error of the estimation is less than $\pm 20\,\mu$m.

7.6   Conclusions

An enhanced accuracy OFDM Radar is presented in this chapter, in which the FRFT and phase analysis techniques are adopted. FRFT is used to achieve the ranging accuracy of the target within one wavelength of the millimeter wave, followed by the utilization of the carrier phase analysis method to further improve the ranging accuracy, which can make the ranging accuracy of the target reach $20\,\mu$m.

Compared with the other published works summarized in Table 7.1, the proposed OFDM radar can achieve high accuracy at a high refresh rate over a long unambiguous range which mainly benefits from the parallel characteristics of OFDM and the short wavelength of a millimeter wave. In future work, we will adopt a higher frequency such as Terahertz band to further improve range accuracy. In addition, multiple millimeter wave OFDM radar sensors can be used for multidimensional ranging for high-precision positioning and imaging.

References

1. R. Drath, A. Horch, Industrie 40: Hit or hype? [industry forum]. IEEE Indus. Electron. Mag. **8**, 56–58 (2014)
2. C.I.S. Han, Z. Xu, S. Wang, Q. Sun, Y. Chen, New paradigm of 5g wireless internet. IEEE J. Sel. Areas Commun. **34**, 474–482 (2016)

3. M. Khoshnevisan, V. Joseph, P. Gupta, F. Meshkati, R. Prakash, P. Tinnakornsrisuphap, 5g industrial networks with comp for urllc and time sensitive network architecture. IEEE J. Sel. Areas Commun. **37**, 947–959 (2019)

4. S.A.K. Tanoli, Utilizing 5g spectrum for healthcare to detect the tremors and breathing activity for multiple sclerosis. Trans. Emerg. Telecommun. Technol. **29**(7), e3454 (2018)

5. D. Haider, A. Ren, D. Fan, N. Zhao, X. Yang, S.A. Shah, F. Hu, Q.H. Abbasi, An efficient monitoring of eclamptic seizures in wireless sensors networks. Comput. Electr. Eng. **75**, 16–30 (2019)

6. H. Korber, H. Wattar, G. Scholl, Modular wireless real-time sensor/actuator network for factory automation applications. IEEE Trans. Indus. Inform. **3**, 111–119 (2007)

7. K. Islam, W. Shen, X. Wang, Wireless sensor network reliability and security in factory automation: a survey. IEEE Tran. Syst. Man Cybern. Part C (Appl. Rev.) **42**, 1243–1256 (2012)

8. J. Lee, W. Li, J. Shen, C. Chuang, Multi-robotic arms automated production line, in *2018 4th International Conference on Control, Automation and Robotics (ICCAR)*, pp. 26–30 (2018)

9. H.G. Jung, Y.H. Cho, P.J. Yoon, J. Kim, Scanning laser radar-based target position designation for parking aid system. IEEE Trans. Intell. Transp. Syst. **9**, 406–424 (2008)

10. L. Piotrowsky, T. Jaeschke, S. Kuppers, N. Pohl, An unambiguous phase-based algorithm for single-digit micron accuracy distance measurements using fmcw radar, in *2019 IEEE MTT-S International Microwave Symposium (IMS)*, pp. 552–555 (2019)

11. M. Pauli, B. Gottel, S. Scherr, A. Bhutani, S. Ayhan, W. Winkler, T. Zwick, Miniaturized millimeter-wave radar sensor for high-accuracy applications. IEEE Trans. Microwave Theory Techniques **65**, 1707–1715 (2017)

12. S. Scherr, B. Göttel, S. Ayhan, A. Bhutani, M. Pauli, W. Winkler, J.C. Scheytt, T. Zwick, Miniaturized 122 ghz ism band fmcw radar with micrometer accuracy, in *2015 European Radar Conference (EuRAD)*, pp. 277–280 (2015)

13. S. Ayhan, S. Thomas, N. Kong, S. Scherr, M. Pauli, T. Jaeschke, J. Wulfsberg, N. Pohl, T. Zwick, Millimeter-wave radar distance measurements in micro machining, in *2015 IEEE Topical Conference on Wireless Sensors and Sensor Networks (WiSNet)*, pp. 65–68 (2015)

14. F. Barbon, G. Vinci, S. Lindner, R. Weigel, A. Koelpin, A six-port interferometer based micrometer-accuracy displacement and vibration measurement radar, in *2012 IEEE/MTT-S International Microwave Symposium Digest*, pp. 1–3 (2012)

15. S. Lindner, F. Barbon, S. Linz, S. Mann, R. Weigel, A. Koelpin, Distance measurements based on guided wave 24ghz dual tone six-port radar, in *2014 11th European Radar Conference*, pp. 57–60 (2014)

16. C. Sturm, W. Wiesbeck, Waveform design and signal processing aspects for fusion of wireless communications and radar sensing. Proc. IEEE **99**, 1236–1259 (2011)

17. Q. Wu, F. Zhao, X. Ai, X. Ai, J. Liu, S. Xiao, Compressive-sensing-based simultaneous polarimetric hrrp reconstruction with random ofdm pair radar signal. IEEE Access **6**, 37837–37849 (2018)

18. C. Knill, F. Roos, B. Schweizer, D. Schindler, C. Waldschmidt, Random multiplexing for an mimo-ofdm radar with compressed sensing-based reconstruction. IEEE Microwave Wirel. Components Lett. **29**, 300–302 (2019)

19. S. Sen, A. Nehorai, Target detection in clutter using adaptive ofdm radar. IEEE Signal Process. Lett. **16**, 592–595 (2009)

20. S. Sen, Adaptive ofdm radar waveform design for improved micro-doppler estimation. IEEE Sensors J. **14**, 3548–3556 (2014)

21. Y. Liu, G. Liao, J. Xu, Z. Yang, Y. Zhang, Adaptive ofdm integrated radar and communications waveform design based on information theory. IEEE Commun. Lett. **21**, 2174–2177 (2017)

22. J. Lim, S. Kim, D. Shin, Two-step doppler estimation based on intercarrier interference mitigation for ofdm radar. IEEE Antennas Wirel. Propag. Lett. **14**, 1726–1729 (2015)

23. G. Hakobyan, B. Yang, A novel intercarrier-interference free signal processing scheme for ofdm radar. IEEE Trans. Veh. Technol. **67**, 5158–5167 (2018)

24. X. Tian, T. Zhang, Q. Zhang, Z. Song, High accuracy doppler processing with low complexity in ofdm-based radcom systems. IEEE Commun. Lett. **21**, 2618–2621 (2017)

Chapter 8
FTN-Enhanced DFT-s-OFDM System

In this chapter, we investigate the enhancement of PAPR and throughput for DFT-s-OFDM system. Section 8.1 introduces the background and motivation of the designs of waveforms based on DFT-s-OFDM. Section 8.2 illustrates the signal model and performance metrics used in the simulation process. Section 8.3 presents the proposed IOTA-based FTN-DFT-s-OFDM Waveform Design. Section 8.4 shows extensive simulation results of the proposed waveform with respect to PAPR, BER and throughput under various different modulation coding schemes (MCSs). Section 8.5 concludes this chapter.

8.1 Introduction

High frequency, i.e., mmWave and THz wireless communication technologies, which exploits the treasure of large bandwidth in higher frequency, has become a research hotspot with the evolution of B5G and 6G [1–4]. The aim of high frequency technology is to provide extremely high-speed data rate for various applications, such as vehicular networks, satellite backhauls and extended reality (XR) applications, etc. [5–10]. The characteristics of the devices and the propagation environment are significantly different at high frequency. For example, the linear range of the amplifier at high frequency is small, which causes unexpected distortion on the signal waveforms with large variance [11]. The design of waveforms thus becomes a major concern in enabling high frequency communication.

The designs of waveform based on OFDM have been commercialized in 4G/5G, and are the baseline of 6G high frequency communication [12, 13] now. Parallel transmission of high speed data is realized in OFDM through time and frequency division multiplexing, which has good resistance to multi-path fading and can support efficient multi-user access. Discrete Fourier transform (DFT) spreading technique has been incorporated in OFDM, termly DFT-s-OFDM, to achieve lower PAPR so as to make it compatible with the low-ability power amplifier, e.g. cellular uplink and ground-to-satellite link. While maintaining the characteristics of single-carrier waveforms, DFT-s-OFDM offers more flexibility compared to the conventional single-carrier system. DFT-s-OFDM has been applied in high frequency communication scenarios [14], which realize multi-user multiplexing when ensuring good PAPR performance. Also, it has been proved that DFT-s-OFDM can achieve a balance between user scheduling flexibility and computational complexity of channel equalization [15], which enables the usage in wireless hotspots with densely distributed user equipment. Recently, waveform variants based on DFT-s-OFDM, e.g., generalized DFT-s-OFDM which redesigns the CP [16, 17], have triggered more research interests to achieve lower latency and out-of-band emission. Despite the above variances, achieving the improvement of both the PAPR and throughput at the same time is a challenge. To enhance the spectral efficiency, utilizing high-order modulation is a straightforward method. However, the different amplitudes of modulated symbols leads to larger PAPR, which makes reducing PAPR and enhancing throughput contradictory. As for the methods which achieve lower PAPR, e.g., applying DFT spreading on OFDM, they cannot improve the spectral efficiency. PAPR can be further lowered by method such as peak clipping. However, this can cause signal distortion and the spectral efficiency is sacrificed. FTN signaling is an efficient method for enhancing the spectral efficiency, with the disadvantage of low transmission reliability due to the additional ISI brought by acceleration.

In order to further enhance the spectral efficiency of transmission, FTN technology has been put forward and introduced into some existing waveform frameworks. Compared to the systems based on Nyquist criteria, FTN signaling can achieve 25% improvement in symbol rate without sacrificing the error rate performance. Basic ideas of incorporating FTN signaling in OFDM-based waveforms are introduced in [18]. FTN also exhibits high compatibility with different modulations, where [19] validates the capacity and SNR gain of FTN-based low-order modulations, and [20] proves the power gains of FTN-based high-order modulations in multi-carrier system. As for the receiving technology of FTN, [21] investigates the maximum posterior-based detection of FTN signaling, which achieves up to 186% gains in spectral efficiency compared with its counterpart of Nyquist signaling. To reduce the computational complexity of FTN receiver, [22] proposes a deep learning-aied FTN signal detection method for uplink multi-user transmission scenario. Despite its throughput gain, the major drawback of FTN is that the accelerated symbols will introduce additional ISI, which degrades the transmission reliability [23]. The proposed method provides a candidate solution for jointly enhancing the performance of PAPR and throughput. We first increase the spectral efficiency by integrating FTN with the low-PAPR DFT-s-OFDM. At the same time, we design IOTA shaping filter to reduce the ISI. The proposed method is simple but efficient, which can improve the performance of PAPR and throughput simultaneously.

In this chapter, to enhance the performances of PAPR and throughput simultaneously, we consider the incorporation of FTN with DFT-s-OFDM and aim to tackle the problem of additional ISI caused by FTN. Specifically, we propose an FTN-DFT-s-OFDM waveform based on IOTA filter, which can lower the ISI while keeping the spectral efficiency enhanced. Good focusing ability on both time and frequency domain can be found in IOTA filter, which is vital for reducing the ISI and inter-channel interference (ICI) [24]. Besides, experimental results have shown that IOTA filter helps in obtaining higher computational efficiency [25]. By combining the above-mentioned ingredients, the detailed design of the proposed transmission waveform is presented in the following sections. The low ISI characteristic also makes the low-complexity frequency-domain equalization-based method available for reliable detection. Monte Carlo simulation is performed to validate the joint benefits of deploying IOTA and FTN technologies in DFT-s-OFDM can be obtained with respect to PAPR, BER and throughput. MCSs have been simulated, where up to 3.5dB PAPR gain, 50% throughput gain in high SNR region can be achieved by the proposed scheme, compared to conventional waveforms.

8.2 Signal Model and Performance Metrics

In this section, we consider a general waveform framework, as well as the metrics which can evaluate the performance of a given waveform.

8.2.1 General Waveform Framework

Multi-carrier waveform supports parallel transmission, where high-rate data stream is decomposed into multiple parallel low-rate sub-data streams. Each sub-data stream is independently modulated and superimposed to form the transmission signal.

OFDM waveform is a special kind of multi-carrier transmission technology. In an OFDM system, the transceiver filters form the special modulation filters, where the transmitting filter can be realized by IDFT and the receiving filter can be realized by DFT. With the usage of cyclic prefix, OFDM technology solves the problem of multi-path fading with low complexity. The following part describes the general framework of OFDM.

Denote s as the symbol index, l as the sub-carrier index, L as the number of sub-carrier, and n as the time index. Define $d_{s,l} \in \mathbb{C}$ as the source symbol. Then the general waveform framework based on the concept of OFDM can be expressed as

$$x(n) = \sum_{s=-\infty}^{\infty} \sum_{l=0}^{L-1} d_{s,l} \cdot p(n - sL) \cdot e^{j2\pi l F(n-sL)}, \tag{8.1}$$

where $p(n)$ is the prototype filter, and $e^{j2\pi l F(n-sL)}$ is Fourier transformation kernel corresponding to the value of frequency shift in the l-th sub-carrier, where F is the reciprocal of L, i.e., $F = 1/L$. The above-mentioned time and frequency transformation relationship is also illustrated in Fig. 8.1.

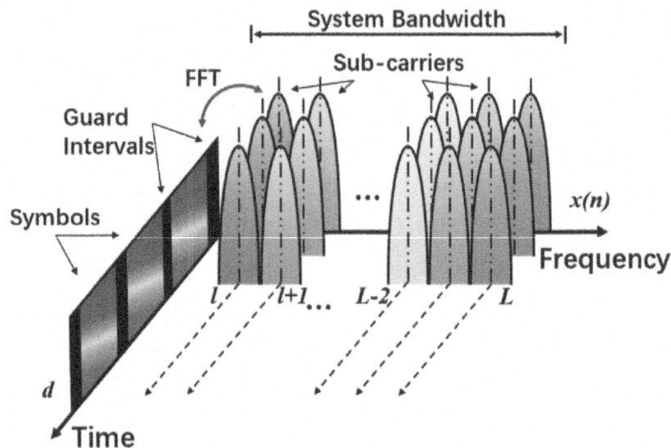

Fig. 8.1 Time & frequency transformation of multi-carrier waveform

8.2.2   *Performance Metrics*

Some major performance metrics of waveforms, i.e., PAPR and throughput, are discussed in the following.

8.2.2.1   Peak-to-Average Power Ratio

Denote $x(n)$ the transmit signal. The performance metric PAPR is defined as the ratio of the peak value to the average value of $x(n)$, which is given by

$$PAPR = \frac{\max|x(n)|^2}{\text{mean }|x(n)|^2}. \tag{8.2}$$

Due to the limited dynamic range of the power amplifier in high frequency transmission, the signal with high PAPR suffers from non-linear distortion. Therefore, PAPR is a key indicator of the power efficiency of the multi-carrier system.

OFDM signal is composed of multiple independently modulated sub-carrier signals. When the phases of the sub-carriers are the same or similar, the superimposed signal will be modulated by the same initial phase signal, resulting in a larger instantaneous power peak value, which further causes a high PAPR. Due to the high PAPR, OFDM signal can easily reach the boundary of the dynamic range of a power amplifier [26]. This can cause unexpected distortion especially in high frequency transmission and may even cause inter-carrier interference due to the leakage. This has hindered the application of OFDM alike waveforms in mmWave and THz communication systems [27, 28].

8.2.2.2   Throughput

Throughput refers to the amount of data that are successfully transmitted per unit of time. The throughput of a transmission system is defined as

$$\boldsymbol{Throughput} = (1 - \mathrm{BLER}) \times \mathrm{TBS}/T, \tag{8.3}$$

where BLER denotes the block error rate, TBS denotes the transport block size and T is the period of transmission data block.

The throughput of OFDM systems depends on the modulation order, channel coding rate and the number of occupied sub-carriers. However, for future-oriented new scenarios, the transmission rate supported by OFDM is still not high enough. In order to provide good performance for complicated application scenarios, waveforms with low PAPR and high spectrum efficiency are required for high-frequency wireless transmission.

8.3 Proposed IOTA-Based FTN-DFT-s-OFDM Waveform Design

We consider improving the spectral efficiency and reducing the PAPR on the basis of OFDM. Based on the commonly deployed DFT spreading technique, we further enhance the spectral efficiency with FTN signaling, where FTN acceleration can reduce the transmission time of each symbol. However, due to the introduction of FTN, ISI is unavoidable. Therefore, the design of prototype filter with better focusing capability is studied.

In Sect. 8.3.1, we present the design of the waveform framework. The corresponding receiving method is illustrated in Sect. 8.3.2.

8.3.1 Transmission Waveform Design

In this part, the components of the proposed transmission waveform are introduced including DFT spreading, FTN signaling and IOTA filter design, respectively.

8.3.1.1 DFT Spreading

DFT spreading can be regarded as a kind of precoding before OFDM modulation. By deploying DFT spreading on the source symbols, the resulted waveform, termly DFT-s-OFDM, can hold the characteristics of single carrier waveform. For the s-th OFDM symbol, we denote $\mathbf{c}_s = \left[c_{s,0}, c_{s,1}, \ldots, c_{s,m}, \ldots, c_{s,M-1}\right]^T$ as the M-dimensional source symbol vector. Then the transformed signal for the s-th symbol duration after M-point DFT is given by

$$d_{s,l} = \frac{1}{\sqrt{M}} \sum_{m=0}^{M-1} c_{s,m} \cdot e^{-j2\pi ml/M}, \quad (l = 0, 1, \ldots, M - 1). \tag{8.4}$$

Substituting (8.4) into (8.1), we have the overall transmit signal for DFT-s-OFDM. The inherent single carrier structure make the above signal a promising enabler of high-frequency communication.

8.3.1.2   FTN Signaling

Given the discrete transmit symbol (8.1), FTN signaling aims to transmit each symbol with a shorter time interval to reserve the time resources. We reuse the above notations, and denote $x(n)$ as the input to the FTN module. Denote T_0 as the normal symbol transmission interval (i.e., the interval under Nyquist sampling), the exact symbol interval after FTN signaling is given by $T = \alpha T_0$, where α is the acceleration factor ranging within $[0, 1]$. The acceleration factor α represents the compression degree of FTN, where the FTN waveform under small α has large spectral efficiency. Denote $g(t - nT)$ as the shaping filter and overall received signal is derived by

$$y(t) = \sqrt{E_s} \sum_n x(n)g(t - nT) + n(t), \tag{8.5}$$

where $n(t)$ denotes AWGN, $g(t - nT)$ denotes the shaping filter, and E_s denotes the average power of the transmitting signal.

FTN can be defined as pulses that are shorter in time and thus no longer orthogonal. We can easily observe that smaller α results in higher data rate. Howevere, due to the non-orthogonality of pulses in FTN signaling, there are overlaps among carriers and ISI appears. The performance of communication systems thus degrade because of the presence of ISI. In the next, we introduce IOTA filter, which helps to resolve the ISI brought by FTN signaling.

8.3.1.3   IOTA Filter Design

IOTA is obtained from orthogonal Gaussian pulse which holds the optimal time-frequency focusing property. IOTA pulse is generally named as root-Nyquist self-transform pulses, where self-transform means that the beginning of the derivation is Gaussian-shaped pulse and root-Nyquist means the end of the derivation is orthogonal pulse.

IOTA filter orthogonalizes the Gaussian filter, and has the same good time-frequency focusing performance as the Gaussian filter [29]. It can be used to reduce the out-of-band emission, PAPR and the ISI brought by FTN signaling.

In the time-domain, the IOTA pulse, denoted as $\xi_{\tau_0}(t)$, can be given by literature [30]

$$\xi_{\tau_0}(t) = \frac{1}{2} \sum_{k=0}^{K-1} \left\{ \overline{d}_{k,v_0} \left[h_{EGF}\left(t + \frac{k}{v_0}\right) + h_{EGF}\left(t - \frac{k}{v_0}\right) \right] \right\} \cdot \sum_{l=0}^{K} \left[\overline{d}_{l,\tau_0} \cos\left(2\pi l \frac{t}{\tau_0}\right) \right],$$
$$-4\tau_0 \le t \le 4\tau_0,$$
$$\tag{8.6}$$

Table 8.1 Parameters of IOTA filter [29]

Parameters	Specifications
τ_0	$1/\sqrt{2}$
v_0	$1/\sqrt{2}$
t	$[-2\sqrt{2}, 2\sqrt{2}]$
K	15
Q	8
$h_{\mathrm{EGF}}(t)$	$= 2^{\frac{1}{4}} e^{-\pi t^2}$

where K is a constant parameter for IOTA, $h_{\mathrm{EGF}}(t)$ represents the extended Gaussian filter (EGF) in the time domain, and $\overline{d}_{l,v_0}$ represents the IOTA coefficients given by

$$\overline{d}_{l,v_0} = \sum_{q=0}^{Q-1} b_{k,q} \cdot e^{-\pi(2q+k)}, 0 \leq k \leq K - 1, 0 \leq q \leq Q - 1, \qquad (8.7)$$

with Q is also a constant value and $b_{k,q}$ is the pre-defined weight coefficients.

The important parameters to generate IOTA filter are listed in Table 8.1.

The proposed waveform uses IOTA filter as the prototype filter by replacing $p(n)$ in (8.1) with (8.6). We get $p(n)$ by discrete sampling on IOTA filter, with the sampling period the same as DFT-s-OFDM (i.e., T/L), where $p(n)$ can be found by

$$p(n) = \frac{1}{2} \sum_{l=0}^{L-1} \left\{ \overline{d}_{l,v_0} \left[h_{EGF}\left(\frac{nt}{L} + \frac{l}{v_0}\right) + h_{EGF}\left(\frac{nt}{L} - \frac{l}{v_0}\right) \right] \right\} \cdot \sum_{l=0}^{L} \left[\overline{d}_{l,\tau_0} \cos\left(2\pi l \frac{nt}{\tau_0 L}\right) \right],$$
$$- 4\tau_0 \leq t \leq 4\tau_0.$$

$$(8.8)$$

IOTA has good performance of focusing on the time and frequency domain, and thus is better on resisting to ISI and ICI than rectangular waveform or root-raised cosine (RRC) filter. IOTA filter also holds similar out of band (OOB) performance as the RRC filter as shown in Fig. 8.2. This indicates that no additional bandwidth is required for deploying IOTA.

8.3.2 Receiver Design

Even with IOTA filter, the existence of ISI still requires a more complicated receiver to recover the original transmission signals. In the sequel, we design the receiving algorithm for the proposed waveform based on frequency-domain equalization (FDE). The receiver design should lay the basis on the conventional DFT-S-OFDM waveform while considering the specific characteristics of FTN and IOTA filter.

Shiya Sugiura [31] proposes a frequency domain equalization (FDE) receiver structure based on MMSE, which can achieve low demodulation complexity, especially for long channel FTN schemes. Specifically, in FDE scheme, a short CP is

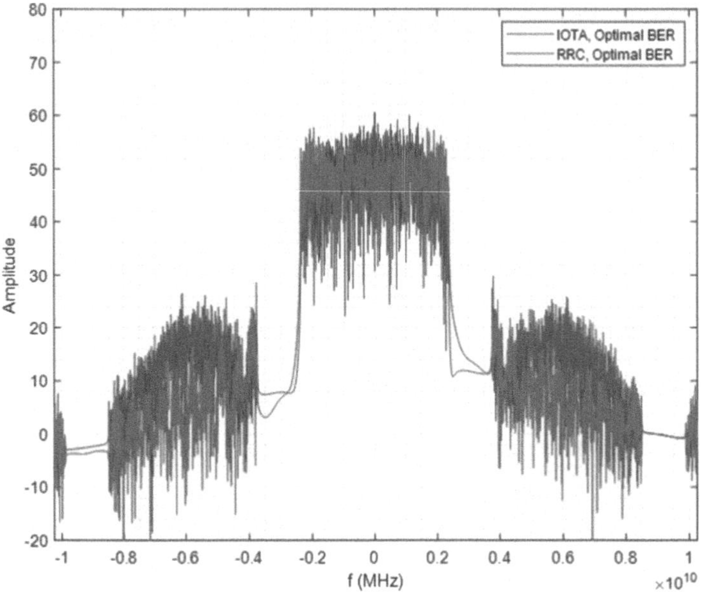

Fig. 8.2 OOB performance of IOTA filter

added to each transport block, and a finite-tap cyclic matrix is adopted to approximate the ISI generated by FTN signaling. Thus, efficient FFT method and a low-
complexity MMSE detection algorithm can be used in the receiver. FDE can achieve
near-optimal BER performance without increasing demodulation complexity and
power consumption.

In AWGN channel, the output signal after matched filter given (8.5) can be
expressed as

$$\hat{y}(t) = y(t) * g(t) = \sqrt{E_s} \sum_n x(n)\hat{g}(t - n\tau T) + \eta(t), \tag{8.9}$$

where $\hat{g}(t) = \int g(\tau)g^*(\tau - t)d\tau$, $\eta(t) = \int n(\tau)g^*(\tau - t)d\tau$, and E_s represents the
average power of transmitting signal. Assuming that the time synchronization is
perfect between the transmitter and receiver, the sampled value of the k-th signal at
the receiver can be written as

$$\begin{aligned}
\hat{y}_k &= \hat{y}(k\tau T) \\
&= \sqrt{E_s} \sum_n x(n)\hat{g}(k\tau T - n\tau T) + \eta(k\tau T) \\
&= \sqrt{E_s}x(n)\hat{g}(0) + \sqrt{E_s} \sum_{n \neq k} x(n)\hat{g}(k\tau T - n\tau T) + \eta(k\tau T),
\end{aligned} \tag{8.10}$$

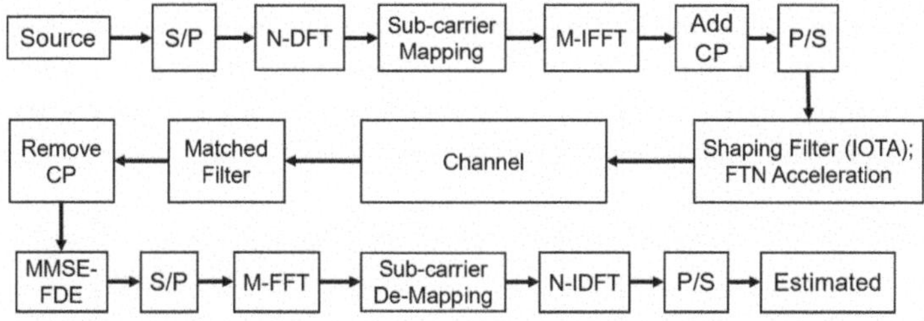

Fig. 8.3 The transceiver design of the proposed waveform

where the first part is the symbol at present, the second part is ISI and $\eta(k\tau T)$ is the zero-mean random Gaussian variable.

Figure 8.3 is the transceiver structure of MMSE-FDE scheme. A CP with length $2v$ is firstly added after N modulated symbols to realize the symbol transmission based on block. Then, the first and last v receiving signal samples are removed from the entire $N + 2v$ samples, and we obtain a received signal block with length N, which can be expressed as

$$
\begin{aligned}
\hat{\mathbf{y}} &= \left[\hat{y}_1, \ldots, \hat{y}_N\right]^T \in \mathbf{C}^N \\
&= \mathbf{G}\mathbf{x} + \boldsymbol{\eta},
\end{aligned}
\tag{8.11}
$$

where $\mathbf{x} = [x_1, x_2, \ldots, x_n]^T$ is the transmitting signal, and $\boldsymbol{\eta} = [\eta_1, \eta_2, \ldots, \eta_N]^T$ is the corresponding channel noise component. The $k - th$ row of tap coefficient matrix $\mathbf{G} \in \mathbf{R}^{N \times N}$ is

$$
\mathbf{G} =
\begin{bmatrix}
g(-v\tau T) & \cdots & g(v\tau T) & 0 & \cdots & \cdots & \cdots & \cdots & 0 \\
0 & g(-v\tau T) & \cdots & g(v\tau T) & 0 & \cdots & \cdots & \cdots & 0 \\
0 & 0 & g(-v\tau T) & \cdots & g(v\tau T) & 0 & \cdots & \cdots & 0 \\
\vdots & \ddots & \ddots & \ddots & \ddots & \ddots & \ddots & \ddots & \vdots \\
0 & 0 & 0 & \cdots & 0 & g(-v\tau T) & \cdots & g(v\tau T) & 0 \\
g((2v-1)\tau T) & \cdots & g(v\tau T) & 0 & \cdots & 0 & g(-v\tau T) & \cdots & g((2v-2)\tau T) \\
g((2v-2)\tau T) & \cdots & g(v\tau T) & 0 & \cdots & 0 & g(-v\tau T) & \cdots & g((2v-3)\tau T) \\
\vdots & \ddots & \ddots & \ddots & \ddots & \ddots & \ddots & \ddots & \vdots \\
g(0) & \cdots & g(v\tau T) & 0 & \cdots & 0 & g(-v\tau T) & \cdots & g(-\tau T)
\end{bmatrix}.
\tag{8.12}
$$

Due to the cyclic structure of matrix $\mathbf{G}$, singular value decomposition can be done as $\mathbf{G} = \mathbf{Q}^T \boldsymbol{\Lambda} \mathbf{Q}^*$, with $\mathbf{Q} \in \mathbf{C}^{N \times N}$ being the eigenmatrix of $\mathbf{G}$, $\boldsymbol{\Lambda}$ being a diagonal matrix and elements in the $i - th$ row being the corresponding FFT coefficients. DFT vector can be used to derive the element in the $l - th$ row and $k - th$ column of $\mathbf{Q}$.

To reduce the computational complexity, the considered FDE algorithm aims to equalize the time-domain receiving signal in the transform-domain with the eigenmatrix $\mathbf{Q}$, which is derived as

$$
\hat{\mathbf{y}}_f \simeq \mathbf{Q}^* \hat{\mathbf{y}} = \boldsymbol{\Lambda} \mathbf{Q}^* \mathbf{x} + \mathbf{Q}^* \mathbf{n} = \boldsymbol{\Lambda} \mathbf{x}_f + \mathbf{n}_f,
\tag{8.13}
$$

where $\mathbf{x}_f$ and $\mathbf{n}_f$ refer to the transformed signal and noise vector, respectively. Then the equalization matrix $\mathbf{W} \in \mathbf{C}^{N \times N}$ is derived based on $\mathbf{\Lambda}$ according to the MMSE criteria, with $\omega_{(i,i)}$ and $\lambda_{(i,i)}$ the $i-th$ rows of the diagonal matrix $\mathbf{W}$ and $\mathbf{\Lambda}$, respectively. Here, $\omega_{(i,i)}$ can be expressed as

$$\omega_{(i,i)} = \lambda_{(i,i)}^* / \left(\left| \lambda_{(i,i)} \right|^2 + N_0 \right). \tag{8.14}$$

Finally, we get the estimated value $\hat{\mathbf{x}}$ of the transmit symbol $\mathbf{x} = [x_1, x_2, \ldots, x_n]^T$ as

$$\hat{\mathbf{x}} = \mathbf{Q}^T \mathbf{W} \hat{\mathbf{y}}_f = \mathbf{Q}^T \mathbf{W} \left(\mathbf{\Lambda} \mathbf{x}_f + \mathbf{n}_f \right). \tag{8.15}$$

The computational complexity of the FDE receiver includes three parts: FFT complexity, weight calculation complexity, and MMSE algorithm complexity. Equation (8.13) uses FFT to transform the received signal into the frequency domain, with its complexity of order $N \log N$. The calculation of the weight in (8.14) requires $4N$ multiplications and the MMSE operation in (8.15) requires $2N$ multiplications, whose complexity are both of order N. Thus, the proposed receiver has the overall computational complexity of $O(N \log N + N)$. The complexity of FDE receiver is only related to the receiving block length N, but not to the FTN acceleration factor α, filter type, or tap length. Therefore, the receiver complexity of the IOTA-based scheme proposed in this chapter is the same as that of the RRC-based non-orthogonal waveform (NOW) scheme. Compared with the time-domain equalization algorithm, whose complexity exponentially increases with tap length, the FDE receiving algorithm used in this chapter has lower complexity, especially under high ISI.

Till now, we have finished the signal demodulation and successfully cancelled the ISI. What is worth noticing is that FDE has relatively low complexity and can similarly achieve the detecting performance of optimal receiver.

8.4   Simulation Results and Analysis

In this section, we implement the proposed IOTA-based FTN-DFT-s-OFDM and evaluate the PAPR, BER, and throughput performance under various MCS settings and acceleration rates. Link-level Monte-Carlo simulations are conducted to valuate the performance gain of the proposed scheme over DFT-s-OFDM and NOW [32], which adopts RRC pulse as the prototype filter of FTN waveform. The main simulation parameters are presented in Table 8.2.

8.4.1   PAPR

In this section, PAPR performance is compared within different MCSs, each with different values of acceleration factor α.

Table 8.2 Parameters for simulation

Carrier frequency	70 GHz
Subcarrier spacing (SCS)	$\Delta f = 960\,\text{kHz}$
Symbol interval	$T = 1.04\,\mu\text{s}$
System bandwidth	800 MHz
DFT size	7792
Guard band	40 MHz
Allocated subcarriers	$M = 792$ (DFT size)
FFT size	1024
Roll-off factor of RRC	$\beta = 1/9$
RRC taps	$20(\alpha = 1)/28(\alpha = 0.95)/30(\alpha = 0.9)/$ $32(\alpha = 0.83)/34(\alpha = 0.77)/42(\alpha = 0.64)/$ $52(\alpha = 0.5)/68(\alpha = 0.38)/102(\alpha = 0.25)$
Up-sampling factor for signal	$K' = \alpha K_0\ (K_0 = 78)$
Channel	AWGN
Channel coding	LDPC
Modulation	QPSK, 16QAM, 64QAM
Baseline Waveforms	DFT-s-OFDM/NOW [32]

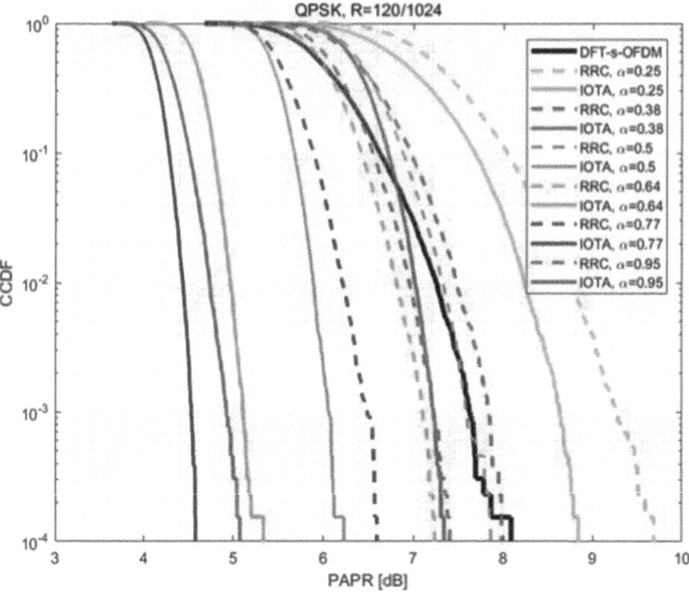

Fig. 8.4 Performance of PAPR based on QPSK with $R = 120/1024$

Figures 8.4, 8.5 and 8.6 show the PAPR performance based on QPSK under various MCSs. From the figures, we can conclude that no matter what filter is used (i.e., IOTA or RRC), the PAPR of FTN first decreases and then increases with the decrease of the acceleration factor α, and the PAPR of FTN reaches the optimal value when the

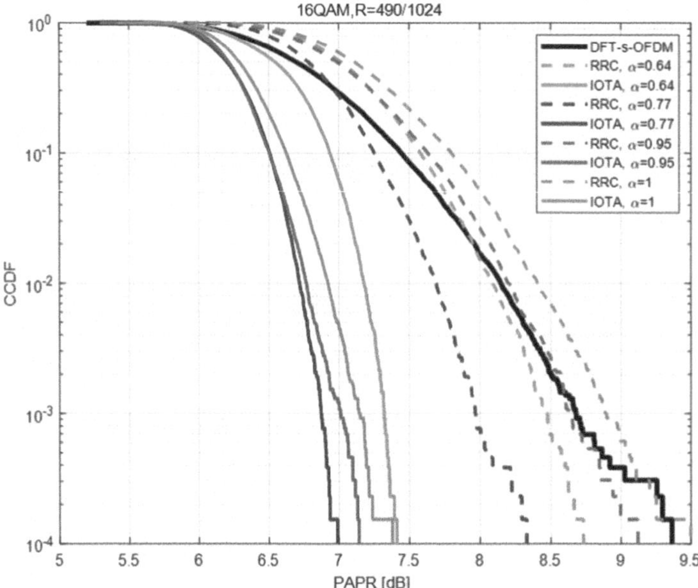

Fig. 8.5 Performance of PAPR based on 16QAM with $R = 490/1024$

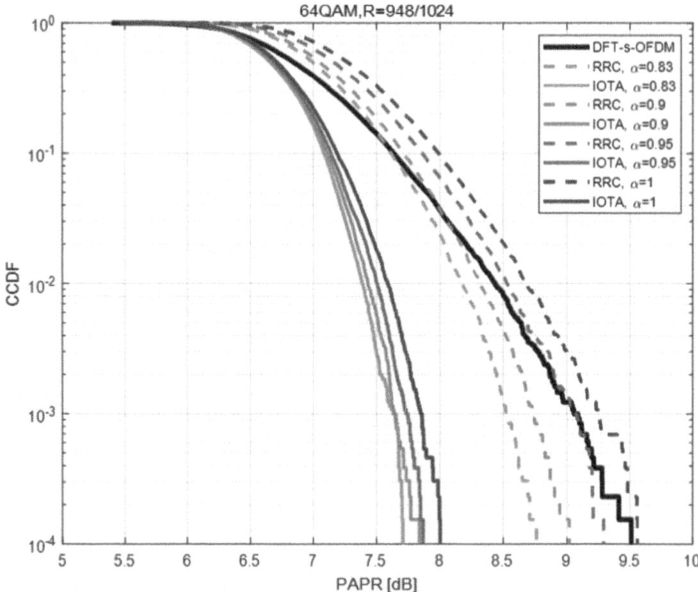

Fig. 8.6 Performance of PAPR based on 64QAM with $R = 948/1024$

acceleration factor is about 0.75-0.85. For higher modulation order, higher spectral efficiency can be expected, and thus higher PAPR value can be seen. However, PAPR performance has little relation with the coding rate R when modulation orders are the same.

Besides, in the simulation process we have totally 9 MCS, each with 9 indexes (i.e., we have 81 cases actually). The corresponding modulation order, coding rate and spectral efficiency are listed in Table 8.3, including the corresponding value of α with experimentally best PAPR performance and PAPR gains based on IOTA in each case. Denoted modulation order as Mod, coding rate as R, spectral efficiency as SE, α for best PAPR on IOTA as best α-IOTA, and α for best PAPR on RRC as best α-RRC.

The reason why the proposed method can reduce PAPR is analyzed in Figs. 8.7, 8.8 and 8.9. When the acceleration factor α of FTN is large, the degree of compression is relatively low. With the increase of compression degree, the average value increases

Table 8.3 PAPR gains under different MCS cases

MCS index	Mod order	R [/1024]	SE	Best α-IOTA	Best α-RRC	PAPR gain over RRC /dB	PAPR gain over DFT /dB
0	2	120	0.2344	0.77	0.77	2	3.5
1	2	379	0.7402	0.77	0.77	2	3.5
2	2	679	1.3262	0.77	0.77	1.8	3.5
3	4	340	1.3281	0.83/0.9	0.77	1.6	2.6
4	4	490	1.9141	0.83	0.77	1.3	2.4
5	4	658	2.5703	0.83	0.83/0.77	1.5	2.5
6	6	438	2.5664	0.83	0.83/0.77	0.9	1.9
7	6	719	4.2129	0.83	0.83	1.3	1.7
8	6	948	5.5547	0.83	0.83	1.1	1.8

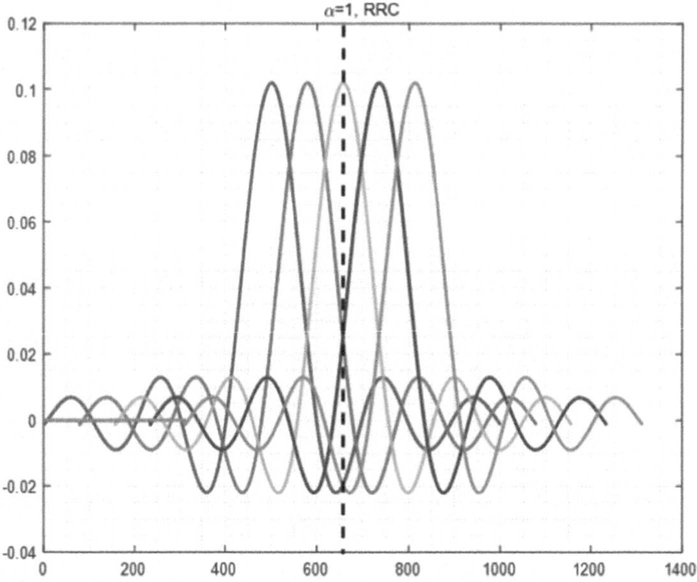

Fig. 8.7 Peak value of RRC filter-based signal with $\alpha = 1$

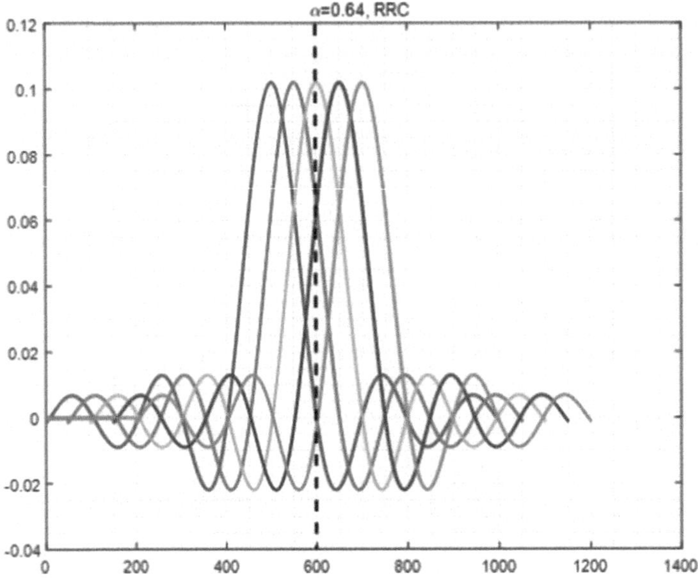

Fig. 8.8   Peak value of RRC filter-based signal with $\alpha = 0.64$

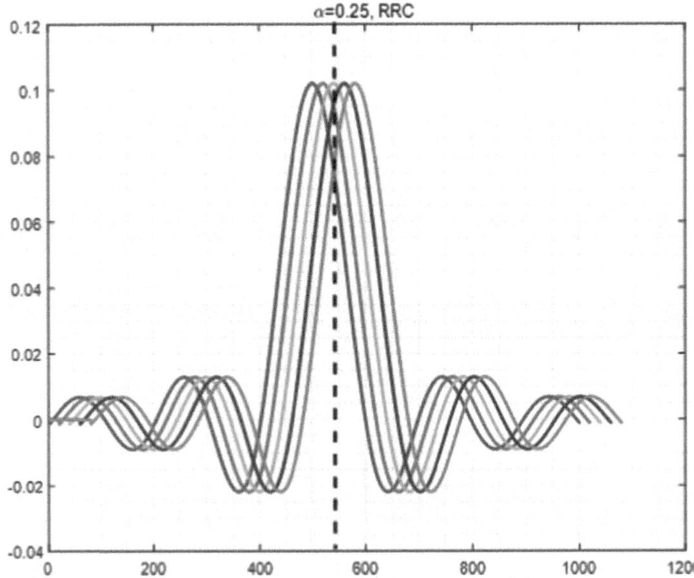

Fig. 8.9   Peak value of RRC filter-based signal with $\alpha = 0.25$

continuously. And the increment of peak value mainly depends on the size of the tails of the pulse. When FTN is further compressed, the size of the peak part is mainly determined by the central envelope of other pulses. In this case, the peak value increases sharply with the increase of the compression degree.

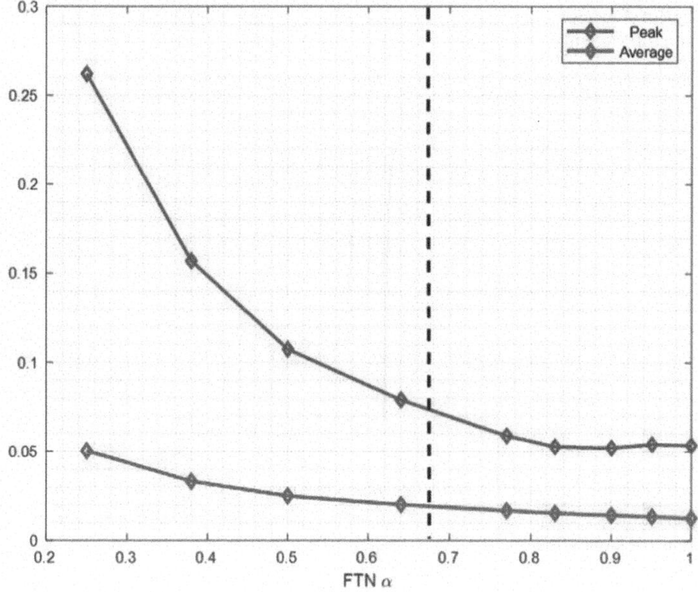

Fig. 8.10 Comparison of peak and average value of FTN under different acceleration factor α

Comparision is made between the peak value and average value of FTN waveform, which is shown in Fig. 8.10. As the FTN acceleration factor α decreases, the average value keeps increasing, whereas the peak value first decreases and then slowly increases and finally increases sharply. Therefore, the PAPR of FTN first decreases and then increases accordingly. This explains the observations of PAPR performance in Figs. 8.4, 8.5 and 8.6.

8.4.2 BER

This part compares the BER performance of IOTA and NOW, based on QPSK and 16QAM with different values of coding rate R and acceleration factor α. NOW denotes the algorithm of FTN-DFT-s-OFDM signaling based on RRC filter.

Figures 8.11 and 8.12 illustrate the BER performance based on the modulation order corresponding to MCS2 and MCS5 in Table 8.3.

From the above two figures, we can derive that the BER performance of IOTA will exceed that of the RRC under small value of α, and also algorithm based on IOTA can support smaller α. When higher spectral efficiency is achieved with higher modulation order, BER performance based on IOTA shows a slower decreasing trend compared with RRC. Thus, we can arrive at the conclusion that IOTA seems to be more suitable for FTN signaling, especially for the case with high spectral efficiency.

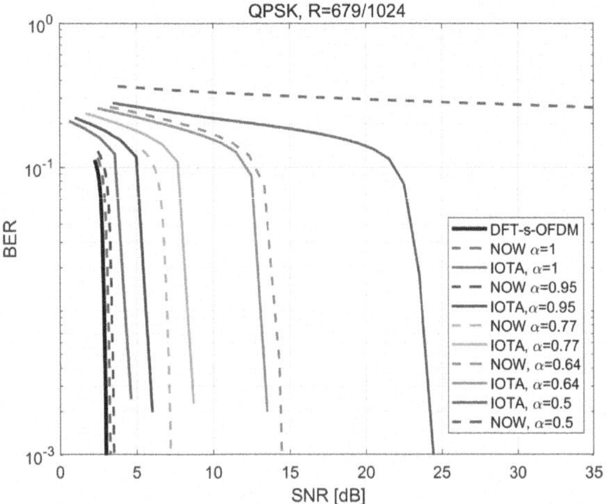

Fig. 8.11 Performance of BER based on QPSK with $R = 679/1024$

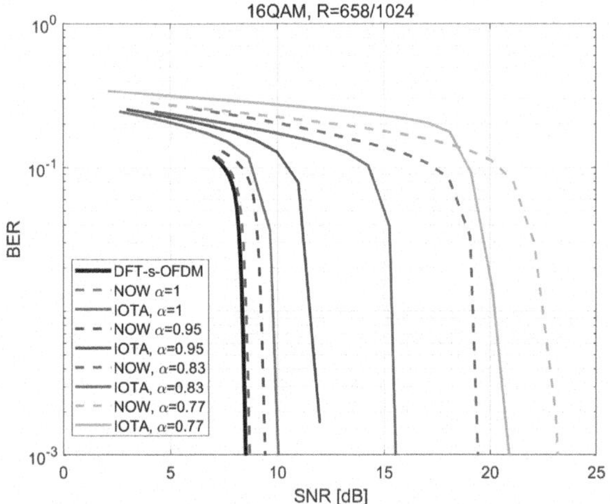

Fig. 8.12 Performance of BER based on 16QAM with $R = 658/1024$

Similarly as PAPR performance mentioned above, we give Table 8.4 listing the BER performance based on IOTA. SNR loss compared to RRC and baseline is listed, as well as the minimum supportable α of IOTA and RRC.

8.4.3   Throughput

In this part, thoughput performance of IOTA and RRC is compared under different α. The throughput is calculated as

$$Troughput = (1 - \mathrm{BLER}) * \mathrm{TBS}/(T1 \times \alpha + T2), \tag{8.16}$$

Table 8.4 Throughput gains under different MCS cases

MCS index	Mod order	R [/1024]	SE	min α-IOTA	min α-RRC	Throughput Gain
0	2	120	0.2344	0.25	0.38	0.52
1	2	379	0.7402	0.38	0.5	0.32
2	2	679	1.3262	0.5	0.64	0.28
3	4	340	1.3281	0.5	0.64	0.28
4	4	490	1.9141	0.64	0.77	0.20
5	4	658	2.5703	0.64	0.83	0.30
6	6	438	2.5664	0.64	0.83	0.30
7	6	719	4.2129	0.77	0.9	0.17
8	6	948	5.5547	0.77	0.9	0.17

where BLER is the block error rate, TBS is the transport block size which represents the number of bits in each block, $T1$ is the transmission period of data block (i.e., slot time), and $T2$ is the transmission period of CP.

Figures 8.13, 8.14 and 8.15 show the throughput performance of systems based on IOTA and RRC filter when different acceleration factors are applied.

By adjusting the MCSs in Table 8.4, we derive the red line in Fig. 8.13 which depicts the envelope of the achievable throughput of RRC-based waveform under different SNRs. For RRC-based waveform, due to the ISI brought by FTN, the

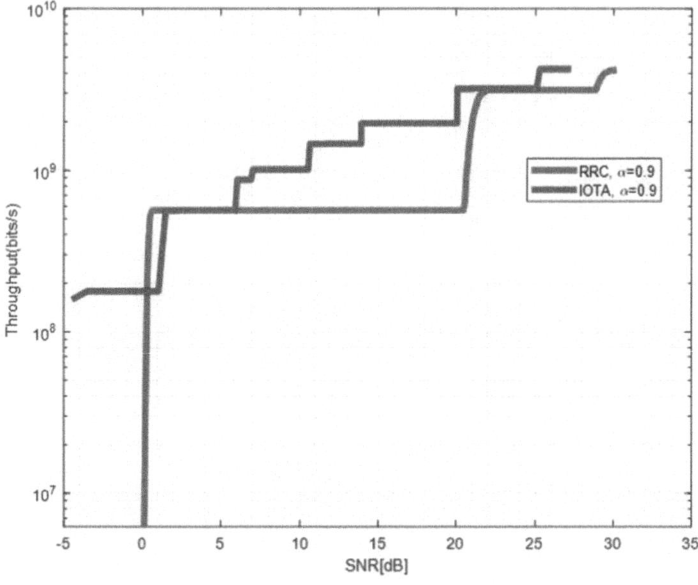

Fig. 8.13 Throughput performance of IOTA and RRC with $\alpha = 0.9$

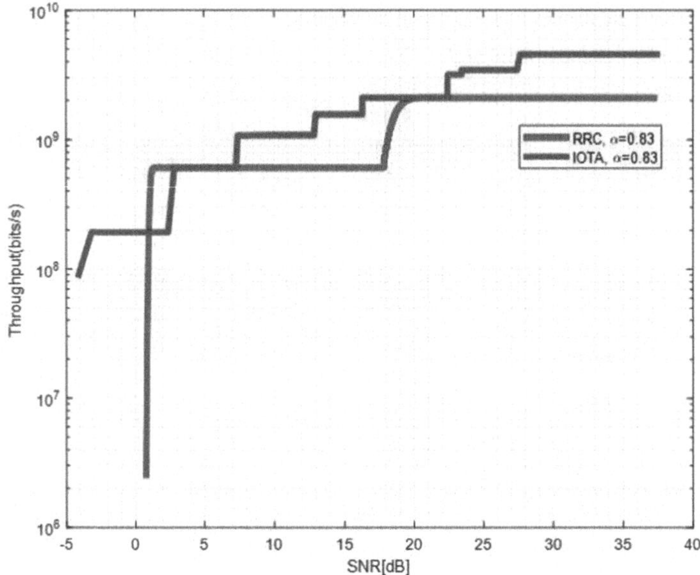

Fig. 8.14 Throughput performance of IOTA and RRC with $\alpha = 0.83$

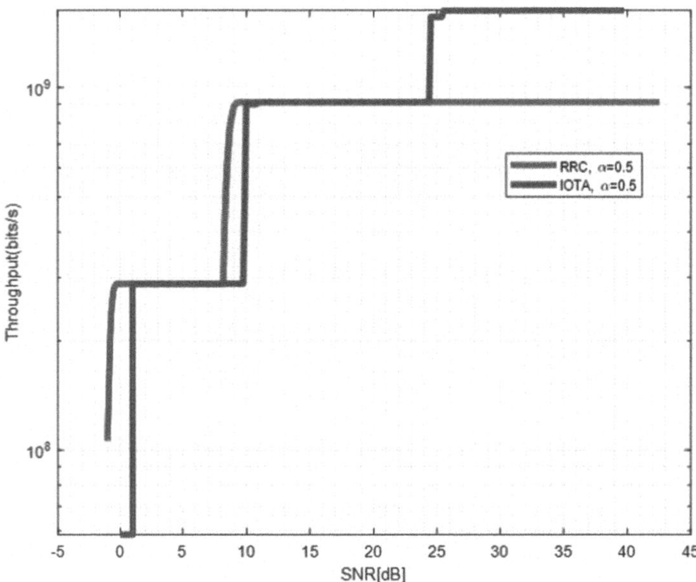

Fig. 8.15 Throughput performance of IOTA and RRC with $\alpha = 0.5$

signal to SINR required for successful decoding of the MCS with a larger index can not be fulfilled even with increased SNR up to 20 dB. This makes the achievable throughput maintain a constant value in a large SNR region. However, for IOTA-based waveform, reduced ISI enhances the SINR and throughput can be enhanced by changing the MCSs. We observe that significant throughput improvement can be seen with SNR between 15-18 dB, which corresponds to the relative SE of MCS 1 and 5 in Table 8.4. Similar observations can be drawn in Figs. 8.14 and 8.15. The throughput performance gain of IOTA is mainly due to its effective ISI reduction.

8.5 Conclusions

In order to meet the new requirements of waveform applied in high frequency communication, where ultra high-speed transmission is one of the vital targets, high spectrum efficiency with reliability is required in the designed waveforms. Based on the OFDM waveform which is basically used in 4G/5G, DFT spreading and FTN signaling are utilized to make the waveform have lower PAPR and higher spectral efficiency, respectively. To further enhance the throughput performance of communication systems, FTN-DFT-s-OFDM waveform based on IOTA filter is proposed in this chapter. The good time-frequency focusing characteristic of IOTA filter enables the ISI reduction and thus the transmission performance of the waveform is improved. On the receiver side, we apply FDE receiver to demodulate the signal, with the complexity of $O(NlogN + N)$. Simulation results have shown that the proposed scheme can offer joint performance gain in terms of PAPR reduction, BER and throughput improvement. In specific, 3.5 dB gain in PAPR and 50% gain in throughput can be achieved compared with the existing waveforms.

For future work, improvements can be further achieved in the waveform design and receiver design. Spectral efficiency and PAPR performance enhancement can be explored based on other waveforms, such as constant envelope OFDM, filter bank of multi-carrier, orthogonal time frequency space, and etc. Moreover, low-complexity iterative design to compact the ISI and ICI is also a promising direction. The state-of-the-art artificial intelligence techniques can also be exploited to enhance the waveform design via an end-to-end fashion [33, 34].

References

1. K. Huq, J.M. Jornet, W.H. Gerstacker, A. Al-Dulaimi, Z. Zhou, J. Aulin, THz communications for mobile heterogeneous networks. THz Commun. Mobile Heterogeneous Networks **56**, 94–95 (2018)
2. W. Hong et al., mmWave 5G NR cellular handset prototype featuring optically invisible beamforming antenna-on-display. IEEE Commun. Mag. **58**, 54–60 (2020)
3. G. Kalfas et al., Next generation fiber-wireless fronthaul for 5G mmWave networks. IEEE Commun. Mag. **57**, 138–144 (2019)

4. K. Aldubaikhy, W. Wu, N. Zhang, N. Cheng, X. Shen, mmWave IEEE 802.11ay for 5G fixed wireless access. IEEE Wirel. Commun. *27*, 88–95 (2020)
5. N. Ye, X. Li, H. Yu, L. Zhao, W. Liu, X. Hou, DeepNOMA: a unified framework for NOMA using deep multi-task learning. IEEE Trans. Wirel. Commun. *19*, 2208–2225 (2020)
6. J. Kim, S.-W. Choi, G. Noh, H. Chung, I. Kim, A study on frequency planning of MN system for 5G vehicular communications. *International Conference on Information and Communication Technology Convergence (ICTC)*, pp. 1442–1445 (2019)
7. Q. Wang, H. Zhang, J.-B. Wang, F. Yang, G.Y. Li, Joint beamforming for integrated Mmwave satellite-terrestrial self-backhauled networks. IEEE Trans. Veh. Technol. **70**, 9103–9117 (2021)
8. R. Torres-Carrión, H. Torres-Carrión, Design and simulation in NS-3 of radio access technology with satellite backhaul for implementation in rural exploitations. *2020 15th Iberian Conference on Information Systems and Technologies (CISTI)*, pp. 1–6 (2020)
9. J. Jung, M. Choi, Y. Goh, J.-M. Chung, Multipath TCP control scheme for low latency and high speed XR real-time M & S devices. IEEE Int. Conf. Consumer Electron. (ICCE) **2022**, 1–3 (2022)
10. J. Pan et al., AI-driven blind signature classification for IoT connectivity: a deep learning approach. IEEE Trans. Wirel. Commun. (2022)
11. J. Li, W. Chen, Q. Zhang, A 600 W broadband doherty power amplifier with improved linearity for wireless communication system. China Commun., 21–29 (2017)
12. R. Zhang, W. Hao, G. Sun, S. Yang, Hybrid precoding design for wideband THz massive MIMO-OFDM systems with beam squint. IEEE Syst. J. **15**, 3925–3928 (2021)
13. N. Ye, J. Yu, A. Wang, R. Zhang, Help from space: grant-free massive access for satellite-based IoT in the 6G era. Digital Commun. Networks **8**, 215–224 (2022)
14. A. Şahin, N. Hosseini, H. Jamal, S.S.M. Hoque, D.W. Matolak, DFT-Spread-OFDM-based chirp transmission. IEEE Commun. Lett. *25*, 902–906 (2021)
15. X. Chen et al., DFT-s-OFDM: enabling flexibility in frequency selectivity and multiuser diversity for 5G. IEEE Consumer Electron. Mag. **9**, 15–22 (2020)
16. G. Berardinelli, K.I. Pedersen, T.B. Sorensen, P. Mogensen, Generalized DFT-spread-OFDM as 5G waveform. IEEE Commun. Mag. **54**, 99–105 (2016)
17. A. Sahin, R. Yang, E. Bala, M.C. Beluri, R.L. Olesen, Flexible DFT-S-OFDM: solutions and challenges. IEEE Commun. Mag. **54**, 106–112 (2016)
18. J.B. Anderson, F. Rusek, V. Öwall, Faster-Than-Nyquist signaling. Proc. IEEE. **101**, 1817–1830 (2013)
19. S. Wen, G. Liu, C. Liu, H. Qu, M. Tian, Y. Chen, Waveform design for high-order QAM Faster-Than-Nyquist transmission in the presence of phase noise. IEEE Trans. Wirel. Commun. **21**, 2–17 (2022)
20. K. Wang, A. Liu, X. Liang, S. Peng, Q. Zhangs, A Faster-Than-Nyquist (FTN)-based multi-carrier system. IEEE Trans. Veh. Technol. **68**, 947–951 (2019)
21. S. Li, B. Bai, J. Zhou, P. Chen, Z. Yu, Reduced-complexity equalization for faster-than-nyquist signaling: new methods based on ungerboeck observation model. IEEE Trans. Commun. **66**, 1190–1204 (2018)
22. J. Pan, N. Ye, A. Wang, X. Li. A Deep learning-aided detection method for FTN-based NOMA. Wirel. Commun. Mobile Comput., 1–11 (2020)
23. S. Li, W. Yuan, J. Yuan, B. Bai, D. Wing Kwan Ng, L. Hanzo, Time-domain vs. frequency-domain equalization for FTN signaling. IEEE Trans. Veh. Technol. **69**, 9174–9179 (2020)
24. R. Razavi, P. Xiao, R. Tafazolli, Information theoretic analysis of OFDM/OQAM with utilized intrinsic interference. IEEE Signal Process. Lett. **22**, 618–622 (2015)
25. M. Xu et al., Orthogonal multiband CAP modulation based on offset-QAM and advanced filter design in spectral efficient MMW RoF systems. J. Lightwave Technol. **35**, 997–1005 (2017)
26. N. Li, M. Li, Z. Deng, A modified Hadamard based SLM without side information for PAPR reduction in OFDM systems. China Commun. **16**, 124–131 (2019)
27. Z. Xing, K. Liu, K. Huang, B. Tang, Y. Liu, Novel PAPR reduction scheme based on continuous nonlinear piecewise companding transform for OFDM systems. China Commun. **17**, 177–192 (2020)

28. J. Bai, Y. Li, W. Cheng, H. Du, Y. Wang, A novel peak-to-average power ratio reduction scheme via tone reservation in OFDM systems. China Commun. **14**, 279–290 (2017)
29. X. Zhou, C. Wang, R. Tang, Channel estimation based on IOTA filter in OFDM/OQPSK and OFDM/OQAM systems[J]. Appl. Sci. **9**, 1454 (2019)
30. P. Siohan, C. Roche, Cosine-modulated filterbanks based on extended Gaussian functions. IEEE Trans. Signal Process. **48**, 3052–3061 (2000)
31. S. Sugiura, Frequency-domain equalization of faster-than-Nyquist signaling. IEEE Wirel. Commun. Lett. **2**, 555–558 (2013)
32. J. Liu, W. Liu, X. Hou, Y. Kishiyama, L. Chen, T. Asai, Non-orthogonal. Waveform, (NOW) for 5G evolution and 6G. *IEEE 31st Annual International Symposium on Personal. Indoor and Mobile Radio Communications*, pp. 1–6 (2020)
33. N. Ye, J. An, J. Yu, Deep learning-enhanced NOMA for massive MTC. IEEE Wirel. Commun. **28**, 66–73 (2021)
34. J. Pan et al., AI-driven blind signature classification for IoT connectivity: a deep learning approach. IEEE Trans. Wirel. Commun., 1 (2022)

Chapter 9
FDSS-Enhanced DFT-s-OFDM System

In this chapter, we present an integrated sensing and communications (ISAC) waveform that adopts FDSS to enhance the performances of DFT-s-OFDM. Section 9.1 introduces the background and motivation of ISAC waveform design. Section 9.2 describes the signal model for the ISAC system and the performance indicators. Section 9.3 shows the OFDM and DFT-s-OFDM sensing waveform, followed by the sensing performance comparison. Section 9.4 presents the FDSS-enhanced DFT-s-OFDM waveform for ISAC, where two FDSS filters and three corresponding detection algorithms are considered. Section 9.5 shows the simulation results. Section 9.6 concludes this chapter.

9.1 Introduction

ISAC systems can perceive the physical characteristics of surroundings while transmitting communication information, which realizes the mutual enhancement of communication and sensing [1]. Approaching 6G, many new services continue to emerge, such as smart cities, smart homes, and autonomous driving. These intelligent services put forward more demands for the integration of sensing and communication [2–4]. Moreover, in 6G, wireless communications and sensing will have more similarities in operating spectrum, hardware structure, and signal processing [5], which provides a further opportunity to integrate the communication and sensing. ISAC will be the emerging technology of 6G, which is characterized by multi-dimensional awareness and ubiquitous communication capability.

The ISAC waveform is an important part in the research of ISAC. It can realize the compatibility between sensing and communication, as well as the efficient utilization of resources. Owing to its multi-carrier characteristics and high spectrum utilization [6, 7], OFDM has been regarded as a suitable ISAC waveform, which can directly

© The Author(s), under exclusive license to Springer Nature Singapore Pte Ltd. 2025   155
J. Li et al., *Key Technologies of High Frequency Wireless Communications*,
https://doi.org/10.1007/978-981-96-5894-7_9

add the sensing function to carry out the sensing detection process. There are considerable researches for ISAC waveform based on OFDM, for example, the robust ISAC waveform based on OFDM inspired by information theory [8], an OFDM chirp waveform diversity for co-designed radar-communication system [9], and ISAC waveform based on the combination of phase coding and OFDM [10, 11].

However, the main challenge of ISAC waveform based on OFDM lies in its high PAPR, which will cause nonlinear distortion and reduce the sensing range [12]. The concept of constant envelope orthogonal frequency-division multiplexing (CE-OFDM) is proposed to reduce the PAPR of OFDM signal [13]. The idea is to implant the OFDM signal into the phase of constant envelope signal, but it will sacrifice a lot of spectral efficiency. Another better method to solve the problem is the combination of DFT expansion with OFDM. By DFT expansion, OFDM has single-carrier characteristics which can reduce the PAPR. DFT-s-OFDM is more suitable for broadband transmission [14], as well as has higher power-amplification efficiency to reduce nonlinear distortion and improve the sensing range, which make it a better candidate for 6G sensing.

Unfortunately, the introduction of DFT will affect the sensing accuracy of OFDM. A lot of effort has been paid to optimizing the sensing performance. From the aspect of transceiver algorithm, Refs. [15, 16] proposed an approximated maximum likelihood radar parameter estimation algorithm to optimize the sensing performance for ISAC system based on OFDM. However, this will rise complexity on the transceiver, which is not suitable for small and intensive equipments. As for the waveform design, a multi-frequency complementary phase-encoded OFDM signal is proposed to promote the ambiguity function property by adjusting the phases of multiple symbols [17]. However, the communication information can only be modulated on the whole phase of OFDM. Subsequently, a direct-sequence spread-spectrum modulation method is proposed to optimize the ambiguity function of the overall signal [18, 19]. Although this method can obtain good ambiguity function performance, the trade-off between communication and sensing performance is not clear.

All together, most schemes simply change the signal distribution by encoding or modulating the communication information, and rarely utilize the filter design. Time–frequency domain filters, such as FDSS [20], prototype filters, and shaping filters, can effectively improve the performance of the waveform by changing the time–frequency domain characteristics, and can be flexibly designed for specific sensing or communication properties. For example, the schemes of pruned DFT-spread filter bank multicarrier (FBMC) [21, 22], FFT-based spectrally efficient frequency-division multiplexing (SEFDM) [23, 24], and more compact FTN signals [25] have different tradeoffs between PAPR, BER, and spectrum efficiency (SE). Frequency-domain spectral shaping on DFT-s-OFDM for SE and PAPR performance has been proposed [26, 27]. Exploiting FDSS, a framework for chirp-based communications, is proposed in [28]. However, it is rare that works focus on the optimization of sensing performance. Hence, utilizing the filter design to jointly optimize communication and sensing performance is worth consideration for ISAC systems.

Finally, it should be noted that alternative modulation or physical-layer approaches, such as FBMC, SEFDM, and FTN signals, have been proposed in the literature, with different tradeoffs between spectrum localization, resource allocation flexibility, multiantenna support, power-efficiency, and transceiver processing complexity. However, in most cases, these methods would cause high system complexity and call for a larger redesign of the 5G NR physical-layer specifications, and are thus not explicitly considered in our scheme.

In this chapter, we propose an ISAC waveform which adopts FDSS to enhance the sensing performance of DFT-s-OFDM by adjusting the correlation of signals, as well as reduce the PAPR of DFT-s-OFDM. FDSS is a kind of modulation technology which uses a spectrum shaping sequence to effectively shape the waveform in a frequency domain. Specifically, we first establish a signal model for the ISAC system, and illustrate two sensing performance indicators, that is, ambiguity function and estimation error. Then, we investigate the ISAC waveform's structure based on OFDM and DFT-s-OFDM, followed by the sensing performance comparison. Through the comparison, we can conclude that the smooth waveform helps to improve the sensing performance by weakening the effect of noise enhancement. On this basis, a framework of FDSS-enhanced DFT-s-OFDM for ISAC is introduced, and two methods, a pre-equalization filter and IOTA filter, are proposed as the FDSS filters to shape the DFT-s-OFDM waveform. Finally, through the simulation results, we observe that the proposed scheme can obtain about 4 dB performance more gain in terms of sensing accuracy than DFT-s-OFDM and outperform than other shaping methods. Meanwhile, the proposed scheme has the lower ambiguity function sidelobe, which means the enhanced resolution. Further, we study the effect of FDSS on PAPR performance. Compared with DFT-s-OFDM, FDSS can effectively reduce PAPR.

The main contributions of this chapter are summarized as follows:

- We propose an FDSS-enhanced DFT-s-OFDM framework for ISAC to jointly improve the sensing and communication performance. Different from most works that only use FDSS to reduce the communication performance, we aim to enhance the sensing performance of DFT-s-OFDM by adjusting the correlation of signals, as well as reducing the PAPR of DFT-s-OFDM further.
- We propose the design idea of an FDSS filter which improves sensing accuracy by smoothing the signal amplitude and weakening the effect of noise enhancement. Inspired by this idea, we develop an FDSS filter based on pre-equalization of frequency-domain signals. It can significantly reduce the fluctuation of signals so as to weaken the noise enhancement caused by the channel estimation process. This way, the performance in terms of sensing accuracy can be improved. To our best knowledge, no one has proposed such scheme in the previous works.
- We evaluate the performance of the FDSS-enhanced DFT-s-OFDM waveform from both sensing and communication aspects, including range estimation accuracy, ambiguity function performance, and BER and PAPR performance, as well as perform the necessary comparison with other methods to verify the effectiveness of the proposed design.

9.2 System Model

In this section, the signal model for ISAC system is briefly described. Then, the performance indicators for the sensing signal are presented.

9.2.1 Signal Model for Integrated Sensing and Communication

As shown in Fig. 9.1, we consider that the system contains one ISAC transmitter (base station) and P ISAC targets (vehicle), and use the same set of waveform and equipment to realize the integration of sensing and communication. Specifically, the communication and sensing processes share the same transmitter. The receiving processing of communication part is completed at the vehicle, and the receiving processing of the sensing part is completed at the base station.

The ISAC signal $x(t)$ is transmitted from the ISAC transceiver, and P echo signals scatter from targets with different distances R_p, and the corresponding relative velocities $v_p, \forall p = 1, 2, \ldots, P$ are received by the ISAC transceiver. Accordingly, the delay of the target is $\tau_p = 2R_p/c$ and the Doppler shift is $f_{dp} = 2f_c v_p/c$, where c is the speed of light and f_c is the carrier frequency of the transmitted signal. Then, the received echo signal $y_p(t)$ of the object with ditances R_p can be expressed as [15]

$$y_p(t) = x(t - \tau_p)e^{j2\pi f_{dp}t} + n(t) \tag{9.1}$$

where $n(t)$ is additive Gaussian white noise. Through the processing of echo signals, parameters such as time delay and frequency offset are extracted to realize accurate estimation of target attributes such as distance and speed.

Fig. 9.1 ISAC system model

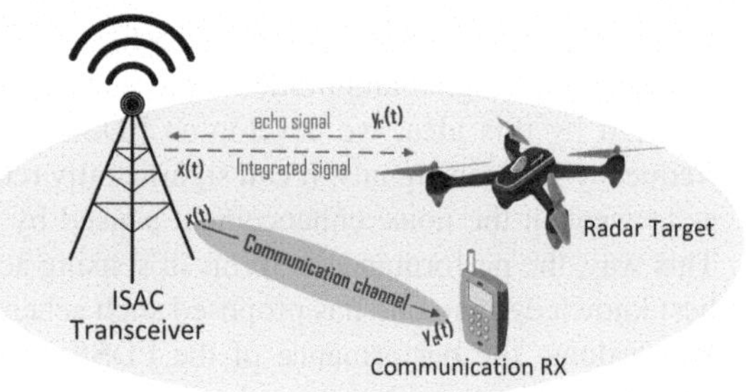

9.2.2 *Performance Indicators*

Some indicators are required to measure the sensing performance of the ISAC system. The performance indicators are introduced as follows, mainly including ambiguity function and estimation error.

9.2.2.1 Ambiguity Function

As an important tool for sensing waveform design and analysis, the ambiguity function can characterize the waveform and corresponding matched filter. When two targets are monitored by the system at the same time, if the positions and velocities of the two objects differ greatly, they can be easily distinguished. However, if the positions and distances of these two objects are very close, it will bring challenges to the system. The ambiguity function is used to measure the ability of the system to distinguish between objectives. By analyzing the ambiguity function of the sensing waveform, we can obtain the resolution, measurement accuracy, and ambiguity of the sensing system under optimal matched filter processing.

There are many ways to define ambiguity functions. The following definition is adopted in this chapter [29].

$$\chi(\tau, f_d) = \int_{-\infty}^{+\infty} y_p(t) y_p^*(t - \tau) e^{i 2\pi f_d t} dt \tag{9.2}$$

where $y_p(t)$ is the echo signal, $y_p^*(t)$ represents the conjugate of $y_p(t)$. τ and f_d are the time delay difference and doppler frequency shift difference between two objects' echo signal, respectively. As $|\chi(\tau, f_d)|$ decreases faster with τ and f_d increasing, the resolution of sensing system is stronger, the ambiguity is lower, and the two objects are easier to distinguish.

9.2.2.2 Estimation Error

Root mean square error (RMSE) is employed as the performance measure for the sensing system, which can be obtained by calculating the standard deviation between the actual distance and the predicted distance. The smaller the value of RMSE, the higher the accuracy of distance estimation. The formula of RMSE is provided as follows.

$$\text{RMSE} = \sqrt{\frac{1}{N} \sum_{i=1}^{n} (Y_i - f(x_i))^2} \tag{9.3}$$

where Y_i is the actual value, such as the distance R_p and the relative velocity v_p of the target object, $f(x_i)$ is the corresponding estimated value, N is the number of data and $\sum$ is the total number of values.

In addition to the above sensing performance, ISAC system also needs to ensure the communication performance, such as spectrum efficiency and bit error rate, as well as considering the power amplifier efficiency, PAPR and other indicators to reduce nonlinear distortion and improve the range of sensing and communication.

9.3   OFDM-Based Waveforms for Sensing

In this section, we mainly introduce the ISAC waveforms structure based on OFDM and DFT-s-OFDM, followed by the sensing performance comparison.

9.3.1   *Vanilla OFDM-Based ISAC Waveform*

The basic principle of OFDM is to divide the high-speed serial data into multiple parts, and each part of the data is carried by a subcarrier. In this way, the symbol rate of the data will be much lower, as well as enhancing the anti-fading and anti-multipath ability of the system. The OFDM signal is transmitted in the form of multiple symbol blocks, which is suitable for the sensing mode.

The traditional integrated OFDM signal pulse consists of one OFDM symbol, which does not consider the transmission of communication information, leading to low communication data rates and difficult synchronization. Therefore, current researchers [15, 16] mostly consider a pulse formed by several consecutive OFDM symbols with communication information, and all OFDM symbols in a pulse form a frame or a complex frame, as to realize the communication function in a pulse.

On this basis, we consider the continuous ISAC waveform based on OFDM symbol as (9.4). Specifically, each sensing pulse consists of N_s OFDM symbols, and each OFDM symbol includes N_c subcarriers. The time duration of one OFDM symbol is T_s, and subcarrier spacing is $\Delta f = 1/T_s$. Then the OFDM-based ISAC signal $x(t)$ can be described as follows [16].

$$x(t) = \sum_{m=0}^{N_s-1} \sum_{n=0}^{N_c-1} a_{m,n} e^{j2\pi n \Delta f(t-mT_s)} \mathrm{rect}(\frac{t-mT_s}{T_s}) \tag{9.4}$$

where $a_{m,n}$ represents the communication information on the n-th subcarrier of the m-th OFDM symbol. The function rect(.) describes a rectangular function, which is equal to 1 for $0 < t \leqslant T_s$, and 0 otherwise. Then, the received echo signal $y_p(t)$ can be expressed as

$$y_p(t) = x(t-\tau_p)e^{j2\pi f_{dp}t} + n(t)$$

$$= \sum_{m=0}^{N_s-1} \sum_{n=0}^{N_c-1} a_{m,n} e^{j2\pi n \Delta f(t-\tau_p-mT_s)} e^{j2\pi m f_{dp}t} \mathrm{rect}(\frac{t-mT_s-\tau_p}{T_s}) + n(t) \tag{9.5}$$

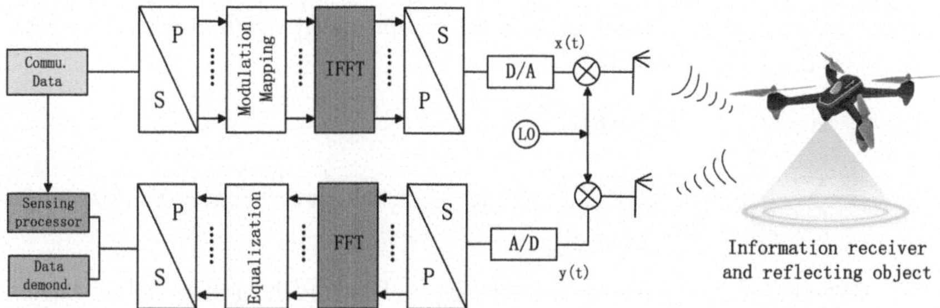

Fig. 9.2 ISAC system structure based on OFDM signal

The ISAC system based on OFDM signal is shown in Fig. 9.2. At the transmitter, the communication data is divided into parallel streams, and mapped onto the modulation symbol sequence. By IFFT, and a subsequent parallel-to-serial conversion, the OFDM-based ISAC signal $x(t)$ is obtained. After the digital to analog conversion, the signal is transmitted. In the receiver, the received modulation symbols are recovered from the received signal $y(t)$ by a FFT operation, while the estimated range and speed information of the sensing target are obtained after sensing processor. The classical detection algorithms mainly includes maximum likelihood estimation algorithm, DFT/IDFT algorithm and so on.

9.3.2 DFT-s-OFDM-Based ISAC Waveform

DFT-s-OFDM and OFDM have similar implementation structures. The main difference is that DFT-s-OFDM has an DFT calculation process before subcarrier mapping, resulting in the difference in signal transmission form. Specifically, DFT-s-OFDM first performs the M-point DFT on the data block. Then the output data is used as the continuous input of OFDM modulator, which is realized by N-point IDFT. Normally $N > M$, the unused inputs of the IDFT are set to zero. Finally, similar to OFDM, it is better to insert CP into each block. On this basis, we consider the ISAC waveform based on DFT-s-OFDM symbol as (9.6).

$$x(t) = \sum_{m=0}^{N_s-1} \sum_{n=0}^{N-1} b_{m,n} e^{j2\pi n \Delta f(t-mT_s)} \mathrm{rect}(\frac{t-mT_s}{T_s}) \tag{9.6}$$

where

$$b_{m,n} = \begin{cases} \sum_{k=0}^{M-1} d_k e^{-j2\pi nk/M}, & 0 \le n < M \\ 0, & M \le n < N \end{cases} \tag{9.7}$$

where d_k represents the communication information [30]. Then, the received echo signal $y_p(t)$ can be rewritten as

$$
\begin{aligned}
y_p(t) &= x(t - \tau_p)e^{j2\pi f_{dp}t} + n(t) \\
&= \sum_{m=0}^{N_s-1} \sum_{n=0}^{N-1} b_{m,n} e^{j2\pi n \Delta f(t-\tau_p-mT_s)} e^{j2\pi m f_{dp}t} \operatorname{rect}(\frac{t - mT_s - \tau_p}{T_s}) + n(t)
\end{aligned}
\tag{9.8}
$$

OFDM modulates the information of the symbol itself to the orthogonal subcarrier, while DFT-s-OFDM modulates the spectrum information of M continuous sampling values in the original sequence to the orthogonal subcarrier. OFDM simultaneously transmits the values of multiple symbols in parallel in a symbol period, and each symbol occupies an independent subchannel, which is equivalent to multiple subchannels transmitting data at the same time. However, DFT-s-OFDM uses all subchannels to transmit the information of the same symbol, which is equivalent to only one single carrier transmitting data, i.e., DFT-s-OFDM signal has the single-carrier nature.

Compared with OFDM, the main advantage of DFT-s-OFDM is the low PAPR, which means the possibility of improving the efficiency of the power amplifier. Simultaneously, DFT-s-OFDM has frequency division multiple access (FDMA) with flexible bandwidth allocation, low complexity, and high-quality equalization in the frequency domain.

9.3.3 Performance Comparisons on OFDM-Based Range Estimation

The sensing performance of the ISAC waveform is mainly considered from sensing range and sensing accuracy. For the sensing range, compared with OFDM, the low PAPR of DFT-s-OFDM can raise the power amplifier efficiency, which is beneficial to improving the sensing range. For the sensing accuracy, we simulate the RMSE for OFDM and DFT-s-OFDM distance estimation under QPSK modulation, using the maximum likelihood estimation algorithm.

As shown in Fig. 9.3, the RMSE for OFDM distance estimation first remains unchanged with the increase of SNR, then falls sharply, and finally decreases linearly, coinciding with CRB. Compared with OFDM, the distance estimation performance of DFT-s-OFDM is 5–10 dB worse. Simulation results show that the sensing accuracy of DFT-s-OFDM is worse than OFDM, and the improvement of DFT-s-OFDM for sensing is necessary. Moreover, under the same configurations, we simulate the RMSE for OFDM distance estimation with different modulation modes. As shown in Fig. 9.4, the distance estimation performance of OFDM under 16 QAM modulation is 5 dB worse than that under QPSK modulation.

Further, the comparison between OFDM signal waveforms under different modulation modes and DFT-s-OFDM signal is shown in Fig. 9.5. Compared with QPSK, the OFDM signal waveform under 16 QAM modulation is more fluctuant, and the

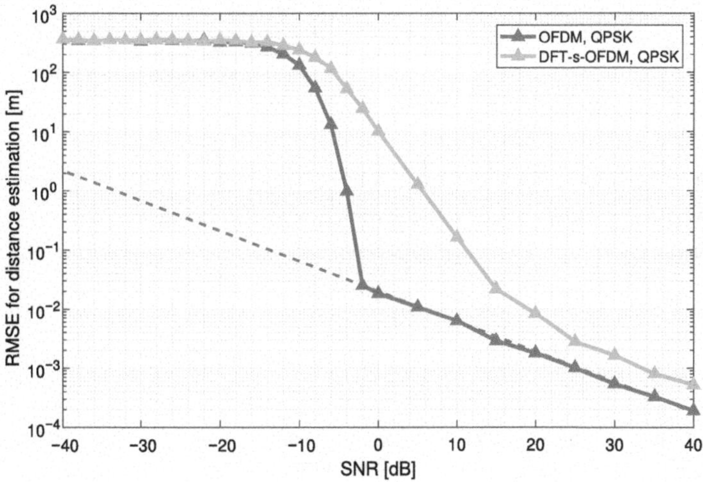

Fig. 9.3 Comparison of the RMSE for distance estimation between OFDM and DFT-s-OFDM

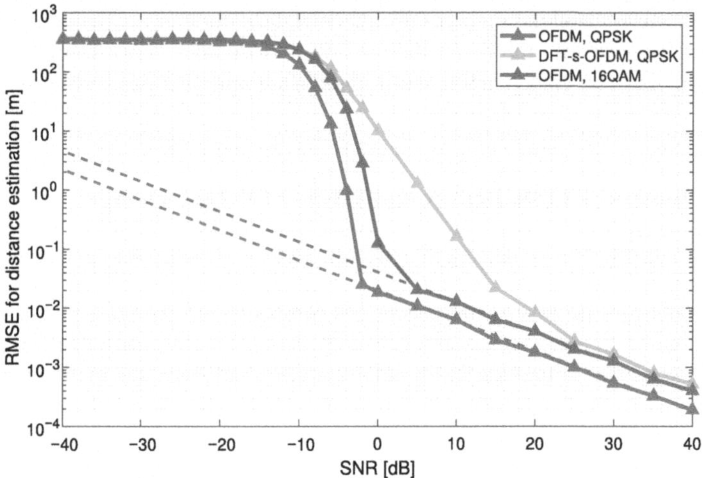

Fig. 9.4 Comparison of the RMSE for distance estimation between 16 QAM and QPSK

DFT-s-OFDM signal has the most serious amplitude fluctuation, i.e., the worst sensing performance.

It is worth noting that, compared with the OFDM signal under QPSK modulation, the amplitude of the OFDM signal under 16 QAM modulation and DFT-s-OFDM signal are not unimodular. Therefore, the noise enhancement for the subcarriers will be caused after the equalization processing at the receiver, where the amplitude of signals are less than 1. Large ripples of the signal waveform degrade the sensing performance due to the noise enhancement with equalization processing.

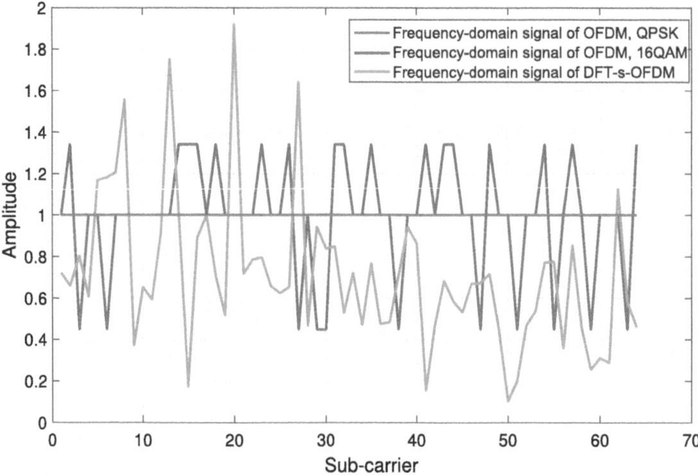

Fig. 9.5 Comparison of OFDM signal under different modulations

From the above analysis emerges the idea that the smoother the frequency domain of the signal waveform, the better the sensing performance. Therefore, the DFT-s-OFDM signal can be shaped in frequency domain to improve its sensing performance.

9.4   Proposed FDSS-Enhanced DFT-s-OFDM Design for Sensing

In this section, we propose a framework for FDSS-enhanced DFT-s-OFDM, which aims to enhance the sensing performance of DFT-s-OFDM by adjusting the relevance of signals, as well as reduce the PAPR of DFT-s-OFDM. Firstly, the FDSS technology is introduced briefly. On the basis, the DFT-s-OFDM signal based FDSS is described, followed by two FDSS filters and three corresponding detection algorithms.

9.4.1   Briefs of FDSS

A set of discrete time domain signals are converted into analog continuous signals after passing through the digital to analog conversion module (DAC). The PAPR of the analog continuous signal has a certain relationship with the correlation between the set of discrete time domain signals [20]. To explain this theory concretely, we propose a set of discrete time domain signals $y(n)$ and a set of discrete data $d(n)$. Then, $yd(n)$ is obtained as

$$y(n) \otimes d(n) = yd(n) \tag{9.9}$$

If $d(n)$ is a set of designed weight coefficient series, the correlation between the adjacent data of $yd(n)$ will be better than that between the adjacent data of $y(n)$. Therefore, the PAPR of $yd(n)$ output signals after DAC will be lower than that of $y(n)$. On this basis, by adjusting the weight coefficient $d(n)$, the correlation between symbols can be adjusted to improve the PAPR performance.

According to the convolution theorem, the convolution operation of two time-domain signals can be equivalent to the multiplication operation of them in the frequency domain. Therefore, by transforming a set of discrete time domain data into discrete frequency domain data after DFT, and then multiplying the data with the designed spectrum shaping sequence, the reshaped sequence can be obtained. Owing to the low complexity of point multiplication operation, this sequence reshaping technology operates better in the frequency domain, which is called FDSS.

In this way, introducing FDSS into signal can adjust the correlation between signals to effectively reduce the PAPR, so as to enhance its communication performance. Simultaneously, FDSS can adjust the waveform amplitude of signals to weaken the impact of noise enhancement, so as to improve the sensing performance.

9.4.2   FDSS-Enhanced DFT-s-OFDM ISAC Waveform

The FDSS-enhanced DFT-s-OFDM symbol can be implemented according to the framework shown in Fig. 9.6. First, operate the M-point DFT on the modulation symbol sequence, i.e., $[d_0, d_1, \ldots, d_{M-1}]$. Then, multiply each term of DFT output sequence by corresponding Fourier coefficient, i.e., FDSS. Finally, operate the N-point IDFT on the shaped sequence padded with zero symbols. Accordingly, the FDSS-enhanced DFT-s-OFDM signal can be described as

$$
p(t) = \sum_{n=0}^{N-1} c_n \underbrace{\sum_{k=0}^{M-1} d_k e^{-j2\pi n \frac{k}{M}}}_{M-\text{point DFT}} e^{j2\pi n \Delta f t} \tag{9.10}
$$

$$
\underbrace{\phantom{\sum_{n=0}^{N-1} c_n \sum_{k=0}^{M-1} d_k e^{-j2\pi n \frac{k}{M}}}}_{\substack{\text{Frequency}-\text{domain spectral shaping} \\ N-\text{point IDFT}}}
$$

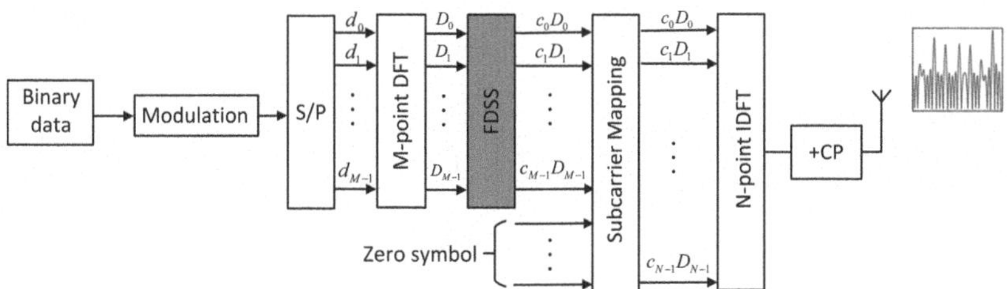

Fig. 9.6 A DFT-s-OFDM transmitter model with a special FDSS filter

where c_n is the weight factor of DFT signal, which can be derived from the Fourier series [28]. By adjusting the form of the Fourier series base, DFT-s-OFDM signals can generate a pseudo-Chip pulse of different shapes, such as linear, triangular, and sinusoidal, which can meet the diverse needs of different sensing scenes. The FDSS-enhanced DFT-s-OFDM ISAC signal $x(t)$ can be rewritten as

$$x(t) = \sum_{m=0}^{N_s-1} \sum_{n=0}^{N-1} d_{m,n} e^{j2\pi n \Delta f(t-mT_s)} \mathrm{rect}(\frac{t-mT_s}{T_s}) \tag{9.11}$$

where,

$$d_{m,n} = \begin{cases} c_n \sum_{k=0}^{M-1} d_k e^{-j2\pi nk/M}, & 0 \le n < M \\ 0 & , M \le n < N \end{cases} \tag{9.12}$$

where d_k represents the communication information. Then, the received echo signal $y_p(t)$ can be rewritten as

$$y_p(t) = x(t - \tau_p) e^{j2\pi f_{dp}t} + n(t)$$
$$= \sum_{m=0}^{N_s-1} \sum_{n=0}^{N-1} d_{m,n} e^{j2\pi n \Delta f(t-\tau_p-mT_s)} e^{j2\pi m f_{dp}t} \mathrm{rect}(\frac{t-mT_s-\tau_p}{T_s}) + n(t)$$
$$\tag{9.13}$$

On the basis of Fig. 9.2, an ISAC system based on FDSS-enhanced DFT-s-OFDM is shown in Fig. 9.7. At the transmitter, the DFT signals are superimposed into a form similar to chirp pulses by weighting the DFT samples, which can realize the sensing and detection functions on the basis of the original communication model. The receiver uses a single-tap MMSE-FDE to remove the impact of the channel. FDSS is applied between DFT and IFFT processes, which is essentially the shaping of frequency domain signals after DFT. Therefore, for the actual receiver, the FDSS coefficient can be regarded as a part of the channel frequency response and estimated through the channel estimation process [28]. In this way, the impact caused by the introduction of FDSS can be eliminated after MMSE-FDE. From this aspect,

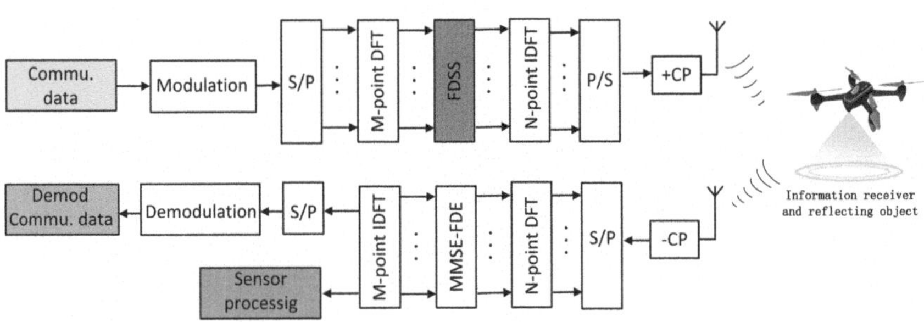

Fig. 9.7 ISAC system structure based on FDSS-enhanced DFT-s-OFDM

a typical DFT-s-OFDM receiver can demodulate the FDSS-enhanced DFT-s-OFDM symbol without any changes. In addition, if the FDSS coefficients are available a priori at the receiver, the receiver can perform better because they do not need to be estimated.

In the following section, we focus on the two important parts of FDSS-enhanced DFT-s-OFDM ISAC system, i.e., the FDSS filter design and the detection algorithm for range estimation.

9.4.3 FDSS Filter Design

Inspired by the comparison between the OFDM sensing performance under different modulation modes, we introduce two implementation methods of FDSS, that is, the pre-equalization filter and the IOTA filter. Both the two FDSS filter design methods make the frequency domain signal amplitude tend to be uniform so as to improve the system sensing performance.

9.4.3.1 Pre-equalization Filter

Frequency domain equalization starts from the frequency domain of the system and directly processes the amplitude frequency characteristics and phase frequency characteristics of the system in the frequency domain, that is, adding frequency domain filters to make them reach the desired transfer function.

In this section, we adopt the pre-equalization filter as a method to realize FDSS. Specifically, we design a filter to equalize the amplitude of the DFT-s-OFDM signal, which aims to change the amplitude of the frequency domain signal close to the average value as much as possible. For each DFT-s-OFDM symbol, we define the frequency domain sequence as $\boldsymbol{m} = \{m_1, m_2, \ldots, m_N\}$, where N is the number of subcarriers, and calculate the average amplitude value $E(\boldsymbol{m})$. Then, taking the average amplitude value as a benchmark, the part beyond the average value is compressed, and the part below the average value is increased to minimize the variance of the symbol amplitude. In this way, we obtain the pre-equalization sequence $\boldsymbol{m}_0$ as

$$\boldsymbol{m}_0 = f(\boldsymbol{m}) = \{f(m_1), f(m_2), \ldots, f(m_N)\},$$

$$\begin{cases} f(m_i) = \xi_a m_i, & m_i > E(\boldsymbol{m}) \\ f(m_i) = \xi_b m_i, & m_i < E(\boldsymbol{m}) \end{cases} \tag{9.14}$$

where ξ_a is the compression ratio, and ξ_b is the increase ratio. The frequency domain amplitudes of DFT-s-OFDM signal and the equalized signal are shown in Fig. 9.8. Compared with the DFT-s-OFDM signal, the frequency domain amplitude of equalized signal becomes more uniform.

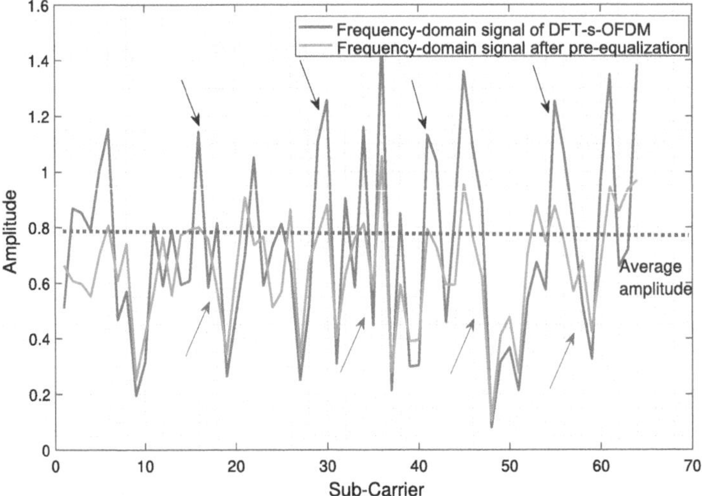

Fig. 9.8 Subcarrier amplitude of DFT-s-OFDM after equalization

9.4.3.2 IOTA Filter

IOTA is a filter function obtained by the time-frequency quadratic orthogonal operation of Gaussian function [31]. Therefore, the IOTA filter has obtained the orthogonality while maintaining the good time-frequency localization (TFL) of Gaussian function. The approximate time-domain IOTA function can be expressed as [32]:

$$\xi_{\tau_0}(t) = \frac{1}{2} \sum_{k=0}^{K-1} \left\{ \bar{d}_{k,v_0} \left[h_{EGF}\left(t + \frac{k}{v_0}\right) + h_{EGF}\left(t - \frac{k}{v_0}\right) \right] \right\} \times \sum_{l=0}^{K} \left[\overline{d}_{l,\tau_0} \cos\left(2\pi l \frac{t}{\tau_0}\right) \right]$$

(9.15)

where the parameter τ_0 and v_0 are chosen to be $1/\sqrt{2}$ normally, and $h_{EGF}(t) = 2^{\frac{1}{4}} e^{-\pi t^2}$ is the EGF in the time domain. $\overline{d}_{k,v_0}$ is the IOTA coefficients, which can be expressed as:

$$\overline{d}_{k,v_0} = \sum_{q=0}^{Q-1} b_{k,q} \times e^{-\pi(2q+k)}, \, 0 \le k \le K - 1, 0 \le q \le Q - 1$$

(9.16)

where K and Q are two parameters of the IOTA filter, generally, $K = 15$ and $Q = 8$.

Figure 9.9 shows the IOTA filter function in the time and frequency domain. It can be observed that the IOTA filter has a relatively ideal TFL and good isotropic properties in the time-frequency domain. Since it has ideal TFL characteristics, IOTA is able to reduce the OOB emission [33, 34], PAPR, and ISI. Therefore, we select the IOTA filter with TFL properties as an FDSS sequence to adjust the correlation between DFT-S-OFDM symbols, so as to improve the sensing performance and

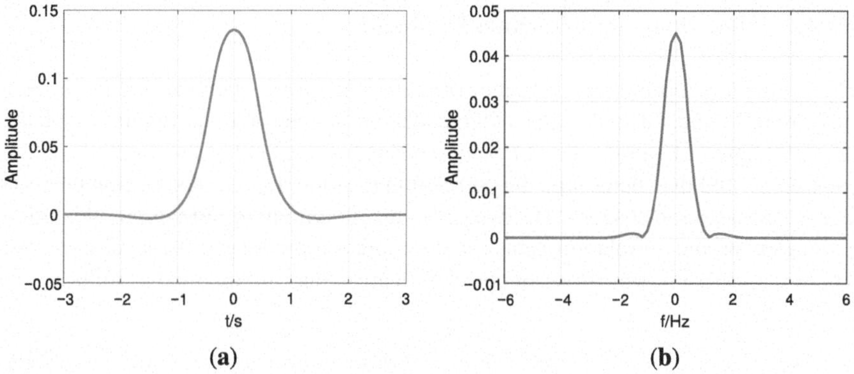

Fig. 9.9   IOTA filter in time and frequency domain. **a** Time domain. **b** Frequency domain

reduce PAPR, while still meeting the OOB emission and pass band signal quality requirements.

9.4.4   Detection Algorithm of FDSS-Enhanced DFT-s-OFDM for Range Estimation

The detection algorithm also affects the sensing performance of the ISAC waveform. Here we consider three corresponding detection algorithms, including DFT detection algorithm, time domain correlation algorithm, and maximum likelihood estimation.

9.4.4.1   DFT Method

Due to the signal characteristics of DFT-s-OFDM, it is easy to have a range estimation by DFT detection. First, the received signal is divided into N_s blocks according to the symbol period. Then, we remove the CP and obtain the m-th received block. At the output of the OFDM demultiplexer, we can obtain the received complex modulation symbols, which contain the range and velocity information of the target object. After element-wise divisions, the range information can be expressed as follows [16].

$$k_r(n) = e^{-j2\pi n \Delta f(2R/c_0)}, 0 \le n \le N - 1 \tag{9.17}$$

where N is the number of subcarriers.

Therefore, the range information is readily obtained by taking an IDFT of $k_r(n)$, which can be expressed as

$$r(k) = \frac{1}{N} \sum_{n=0}^{N-1} e^{-j2\pi n \Delta f(2R/c_0)} e^{j2\pi nk/N} \tag{9.18}$$

9.4.4.2 Time Domain Correlation Estimator

Time domain correlation refers to the correlation processing between the local code, such as PN code, M code, and other codes with good auto-correlation, and the received signal. Through the correlation value we can judge the strength of the correlation, and then determine the correlation position, i.e., the acquisition position and synchronization position. The essence of the time-domain correlation calculation is to calculate the correlation value by summing the product of the local code and received signal. The time-domain correlation algorithm is expressed as follows.

$$\rho(\tau) = \int_{-\infty}^{+\infty} x(t)y(t - \tau)dt \tag{9.19}$$

where $x(t)$ is the transmitted signal, and $y(t - \tau)$ is the received echo signal.

9.4.4.3 Maximum Likelihood Estimator

Maximum likelihood estimation is a widely used parameter- estimation method. The idea of maximum likelihood estimation is that for a given observation data x, we hope to find the parameter θ^* which can generate observation data with maximum probability from all parameters $\theta_1, \theta_2, \ldots, \theta_n$ as the estimation result. The estimated parameter θ^* shall satisfy

$$L\left(\theta^* \mid x\right) = p\left(x \mid \theta^*\right) \geq p(x \mid \theta) = L(\theta \mid x), \theta = \theta_1 \ldots, \theta_n \tag{9.20}$$

In the practical operation, we regard the parameter θ to be estimated as a variable, and calculate the probability function $p(x \mid \theta)$. Then we find the parameter θ which can maximize the probability function to generate the observation data x. The parameter θ is expressed as follows, which can be solved by finding the derivative equal to 0 [15].

$$\theta^* = \arg \max_{\theta} p(x \mid \theta) \tag{9.21}$$

9.5 Simulation Results

To verify the effectiveness of the proposed design, we simulate the performance of FDSS-enhanced DFT-s-OFDM waveform from many aspects, including distance estimation accuracy, ambiguity function performance, and BER and PAPR performance. The simulation framework is according to Fig. 9.7. At the transmitter, the binary communication data are divided into parallel streams, and mapped onto complex-valued QPSK symbols. Then, DFT is performed on the modulation symbol sequence. By weighting the DFT samples with the FDSS coefficient, the DFT sequences are superimposed into a form similar to chirp pulses. After IFFT and a

Table 9.1 Basic parameter configuration

Parameter	Numerical value
Data modulation	QPSK
Carrier frequency	5.89 GHz
Bandwidth	10 MHz
Number of subcarriers	64
Symbol period	6.4 μs
Carrier interval	156.25 kHz
Symbol number	1000
CP number	16
Longest distance	240 m
Antenna gain	100
RCS	1 m^2
Transceiver distance	20 m
Channel model	AWGN
Channel coding	LDPC (code rate 658/1024)

subsequent parallel-to-serial conversion, the FDSS-enhanced DFT-s-OFDM ISAC signal is obtained. Then, time delay and Gaussian noise are added to simulate the transmission process under an ideal channel. In the receiver, the received modulation symbols are recovered from the received signal by a FFT operation. DFT/IDFT algorithm is adopted as the detection algorithm to obtain the range estimation. The simulation is implemented by MATLAB software with a Dell EMC XE2420 Server. Basic parameter configuration is shown in Table 9.1.

9.5.1 *Range Estimation Accuracy for FDSS-Enhanced DFT-s-OFDM*

We simulate the RMSE for the FDSS-enhanced DFT-s-OFDM distance estimation using a pre-equalization filter and IOTA filter. As shown in Fig. 9.10, the RMSE for the equalized DFT-s-OFDM distance estimation is significantly better than that of the baseline. A gain of about 2 dB can be obtained, and even in certain SNR range (14 dB to 26 dB), the gain can reach 4 dB. The simulation result shows that by shaping the DFT-s-OFDM waveform to the average amplitude, the pre-equalization filter can make the amplitude of the equalized DFT-s-OFDM waveform as uniform as possible, and then can improve the sensing performance of DFT-s-OFDM signal.

The RMSE for the DFT-s-OFDM distance estimation under different IOTA is shown in Fig. 9.11, where the waveform of IOTA2 is smoother than that of IOTA1 and performs better. From the results, we can observe that FDSS-enhanced DFT-s-OFDM based on the IOTA filter has the same range sensing performance as DFT-s-OFDM, and can obtain a degree of gain under certain SNR.

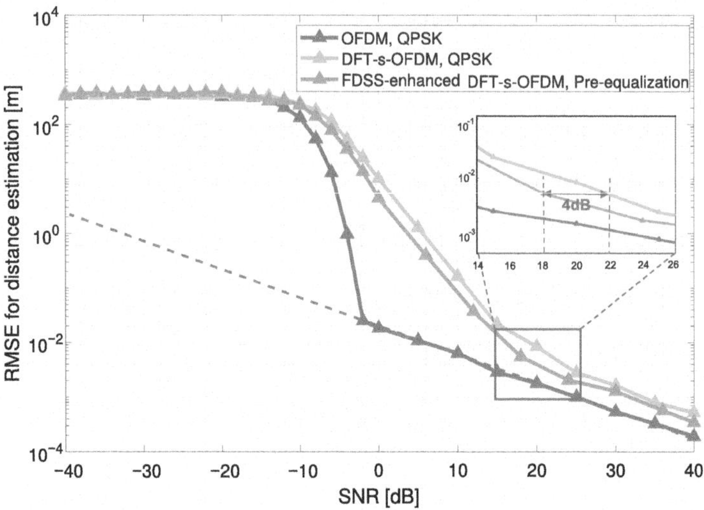

Fig. 9.10 Sensing performance of FDSS-enhanced DFT-s-OFDM with pre-equalization

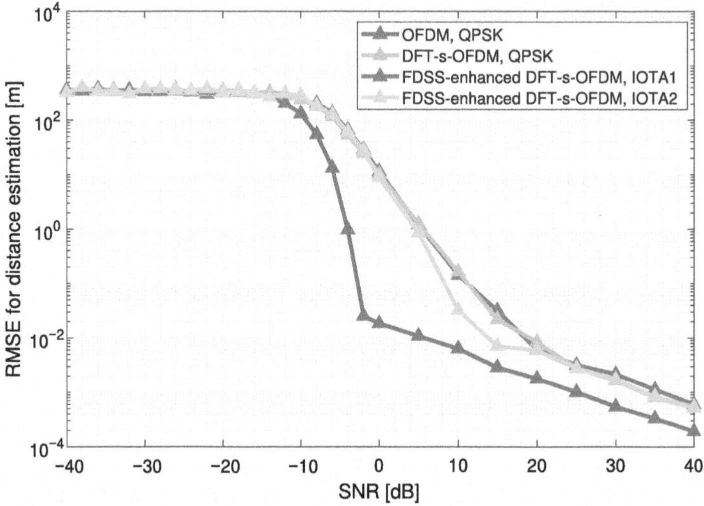

Fig. 9.11 Sensing performance of FDSS-enhanced DFT-s-OFDM with IOTA

We can conclude that the two implementation methods of FDSS can both improve the sensing performance of DFT-s-OFDM waveform. As discussed in Sect. 9.3.3, equalizing the amplitude of the signal in the frequency domain can help to improve the sensing accuracy of the distance, which is reflected in both the pre-equalization and flattening IOTA2.

9.5.2 *Ambiguity Function Performance*

We simulate the ambiguity function to verify the change of FDSS-enhanced-DFT-s-OFDM sensing performance. The steeper the main peak of the ambiguity function, the stronger the resolution of the sensing target. As Figs. 9.12 and 9.13 show, the time-delay domain ambiguity function sidelobe of OFDM is lower than that of DFT-s-OFDM, which means OFDM has better sensing performance. It is consistent with our previous results. After pre-equalization, the ambiguity function of FDSS-enhanced DFT-s-OFDM waveform has a lower sidelobe, which helps to improve the sensing accuracy.

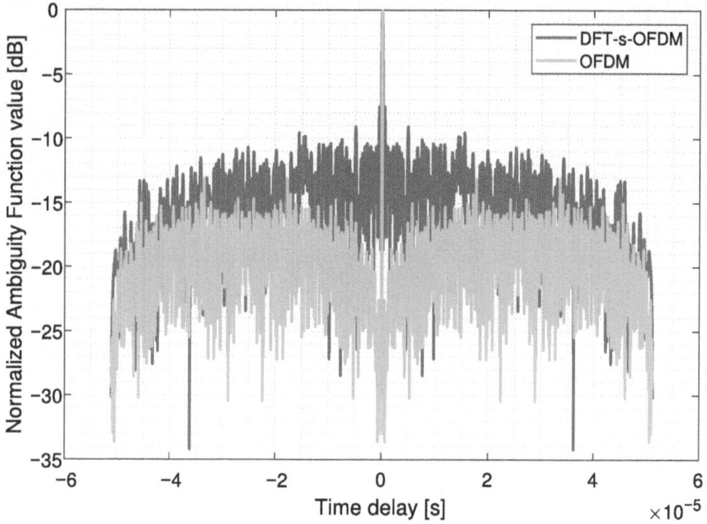

Fig. 9.12 Comparison of ambiguity functions between DFT-s-OFDM and OFDM

Fig. 9.13 Comparison of ambiguity functions between DFT-s-OFDM and FDSS-enhanced DFT-s-OFDM with pre-equalization

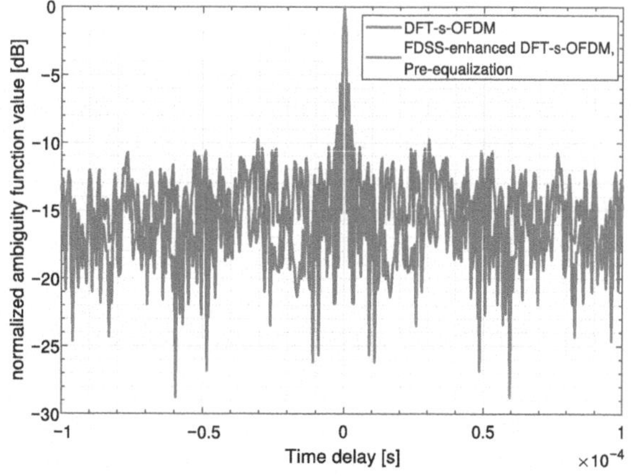

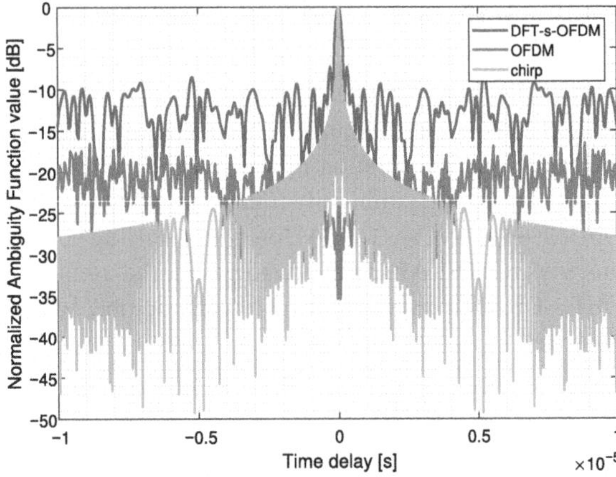

Fig. 9.14 Comparison of ambiguity functions among DFT-s-OFDM, OFDM and chirp

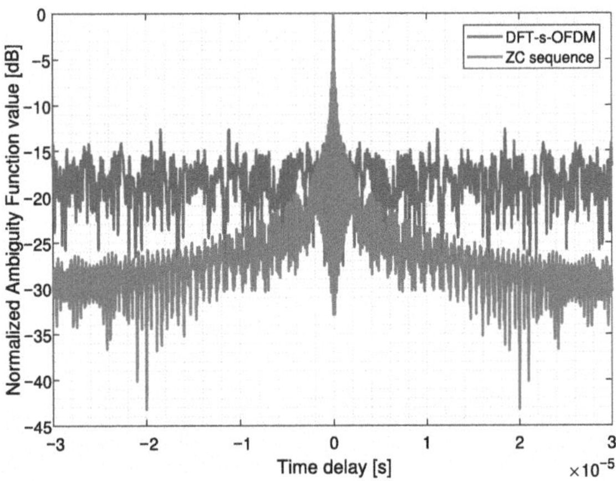

Fig. 9.15 Comparison of ambiguity functions between DFT-s-OFDM and ZC

Moreover, we compare the ambiguity functions between DFT-s-OFDM and other sequences such as chirp and Zadoff–Chu (ZC) sequences. As an ideal sensing waveform, chirp has a lower ambiguity function sidelobe than both OFDM and DFT-s-OFDM, which is shown in Fig. 9.14. For ZC sequences, which have constant amplitude, good autocorrelation and cross-correlation, the ambiguity function sidelobe is lower than DFT-s-OFDM, as shown in Fig. 9.15. Chirp and ZC sequences have constant envelope and average amplitude, so they have better sensing performance, which is similar to our design idea.

9.5.3   BER and PAPR Comparisons

In addition, we simulate the influence of FDSS on PAPR and BER performance. Under the condition of QPSK, we compare the proposed scheme with other FDSS filters such as Bartlett filter, Hann filter and Hamming filter, as well as methods used in other articles such as FTN [26] and raised cosine spectrum shaping [27]. The simulation results of PAPR performance are shown in Fig. 9.16. It can be observed that FTN and RC shaping methods can obtain PAPR gains of 1 dB and 0.2 dB, respectively. The methods based on other filters can obtain PAPR gains of about 2 dB. The FDSS scheme based on IOTA can obtain PAPR gains of 3 dB. Compared with other schemes, FDSS-enhanced DFT-s-OFDM based on an IOTA filter has a better PAPR performance without a loss of sensing performance.

As shown in Fig. 9.17, the introduction of FDSS in DFT-s-OFDM causes a certain degree of loss of BER performance. Under the condition of QPSK, the SNR loss of FDSS scheme based on RC is less than 0.5 dB, and the SNR loss of IOTA is less than 1 dB.

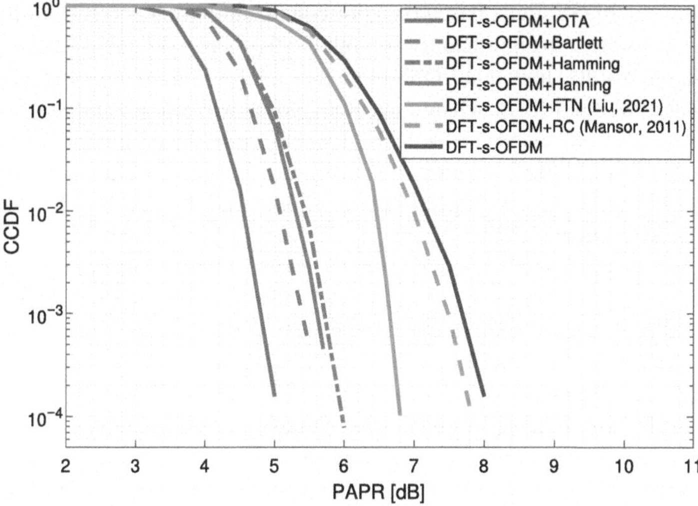

Fig. 9.16 Effect of FDSS on PAPR performance, including the results of FTN [26] and RC [27]

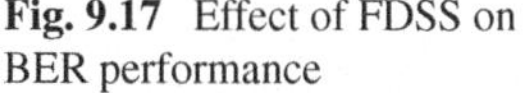

Fig. 9.17 Effect of FDSS on BER performance

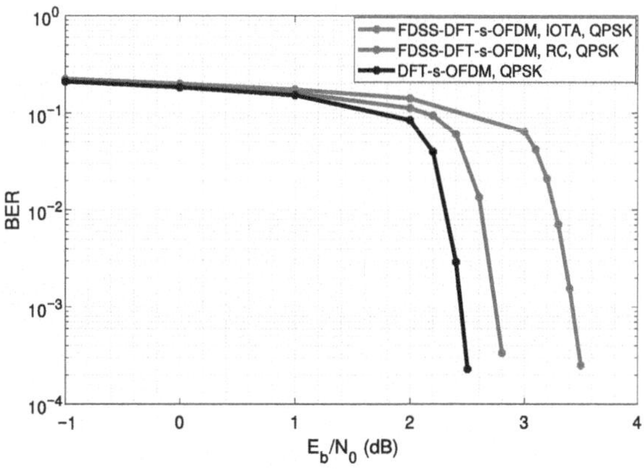

9.6   Conclusions and Future Work

In this chapter, we proposed an FDSS-enhanced DFT-s-OFDM waveform for an ISAC system to improve both sensing and communication performance. Firstly, we established a signal model for the ISAC system, followed by illustrating the corresponding performance indicators. Through the simulation of RMSE for OFDM distance estimation under different modulation methods, we observed that the amplitude fluctuation of the ISAC waveform in frequency domain can affect its sensing performance, owing to the noise enhancement. Specifically, the smoother waveform in the frequency domain causes less noise enhancement, which makes the sensing performance better. On this basis, we proposed a framework of FDSS-enhanced DFT-s-OFDM, which utilized FDSS to shape the DFT-s-OFDM waveform in frequency domain to make the distribution of amplitude more uniform. Simultaneously, we designed the FDSS filters based on a pre-equalization filter and IOTA filter, both of which were shown to significantly smooth the frequency domain signals. Finally, the simulation results show that FDSS-enhanced DFT-s-OFDM can obtain a performance gain of about 4dB in sensing accuracy. Meanwhile, the ambiguity function sidelobe of FDSS-enhanced DFT-s-OFDM waveform is lower than that of DFT-s-OFDM, which means the resolution of sensing is enhanced. Further, we studied the effect of FDSS on signal PAPR performance. Compared with DFT-s-OFDM, FDSS can effectively reduce PAPR.

Moreover, several issues need to be considered in the future. For example, the BER performance of our scheme is not good, and FDSS coefficient setting is not flexible. As a powerful means to realize the complex optimization, deep neural networks (DNN) can be used for more intelligent FDSS coefficient setting, so as to improve the sensing and communication performance jointly for ISAC system.

The main benefit of the proposed design is that it provides a method to effectively improve the performance of ISAC waveform, including sensing accuracy and PAPR property, by changing the time–frequency domain characteristics of the waveform with a filter. In addition, the proposed design can be carried out to meet different sensing or communication performance requirements with high design flexibility, which provides a potential design idea for complex and diverse wireless sensing in 6G application scenarios [35–37].

References

1. A. Liu, Z. Huang, M. Li, Y. Wan, W. Li, T.X. Han, C. Liu, R. Du, D.K.P. Tan, J. Lu et al., A survey on fundamental limits of integrated sensing and communication. IEEE Commun. Surv. Tutor. **24**, 994–1034 (2022). https://doi.org/10.1109/COMST.2022.3149272
2. J. Pan, N. Ye, H. Yu, T. Hong, S. Al-Rubaye, S. Mumtaz, A. Al-Dulaimi, I. Chih-Lin, AI-driven blind signature classification for IoT connectivity: a deep learning approach. IEEE Trans. Wirel. Commun. **21**, 6033–6047 (2022). https://doi.org/10.1109/TWC.2022.3145399

3. N. Ye, X. Li, H. Yu, L. Zhao, W. Liu, X. Hou, DeepNOMA: a unified framework for NOMA using deep multi-task learning. IEEE Trans. Wirel. Commun. **19**, 2208–2225 (2020). https://doi.org/10.1109/TWC.2019.2963185

4. P. Wang, N. Ye, J. Li, B. Di, A. Wang, Asynchronous multi-user detection for code-domain NOMA: expectation propagation over 3D factor-graph. IEEE Trans. Veh. Technol. **71**, 10770–10781 (2022). https://doi.org/10.1109/TVT.2022.3187746

5. T.S. Rappaport, Y. Xing, O. Kanhere, S. Ju, A. Madanayake, S. Mandal, A. Alkhateeb, G.C. Trichopoulos, Wireless communications and applications above 100 GHz: opportunities and challenges for 6G and beyond. IEEE Access **7**, 78729–78757 (2019)

6. B. Nuss, L. Sit, M. Fennel, J. Mayer, T. Mahler, T. Zwick, MIMO OFDM radar system for drone detection, in *Proceedings of the 2017 18th International Radar Symposium (IRS), Prague, Czech Republic*, 28–30 June 2017; pp. 1–9

7. J. Schuerger, D. Garmatyuk, Performance of random OFDM radar signals in deception jamming scenarios, in *Proceedings of the 2009 IEEE Radar Conference*, Pasadena, CA, USA, 4–8 May 2009; pp. 1–6

8. Y. Liu, G. Liao, Z. Yang, Robust OFDM integrated radar and communications waveform design based on information theory. Signal Process. **162**, 317–329 (2019)

9. W.Q. Wang, Z. Zheng, S. Zhang, OFDM chirp waveform diversity for co-designed radar-communication system, in *Proceedings of the 2017 18th International Radar Symposium (IRS)*, Prague, Czech Republic, 28–30 June 2017; pp. 1–9

10. J. Zhao, K. Huo, X. Li, A chaos-based phase-coded OFDM signal for joint radar-communication systems, in *Proceedings of the 12th International Conference on Signal Processing (ICSP)*, Hangzhou, China, 19–23 October 2014; Volume 2015, pp. 1997–2002

11. Y. Yang, J. Mei, D. Hu, Y. Lei, X. Luo, Research on reducing PAPR of QAM-OFDM radar-communication integration sharing signal. J. Eng. **2019**, 8042–8046 (2019)

12. D. Chen, Y. Tian, D. Qu, T. Jiang, OQAM-OFDM for wireless communications in future internet of things: a survey on key technologies and challenges. IEEE Internet Things J. **5**, 3788–3809 (2018)

13. M.A. Dida, H. Hao, X. Wang, T. Ran, Constant envelope chirped OFDM for power-efficient radar communication, in *Proceedings of the 2016 IEEE Information Technology, Networking, Electronic and Automation Control Conference*, Chongqing, China, 20–22 May 2016; pp. 298–301

14. A. Kakkavas, W. Xu, J. Luo, M. Castañeda, J.A. Nossek, On PAPR characteristics of DFT-s-OFDM with geometric and probabilistic constellation shaping, in *Proceedings of the 2017 IEEE 18th International Workshop on Signal Processing Advances in Wireless Communications (SPAWC)*, Sapporo, Japan, 3–6 July 2017; pp. 1–5

15. L. Gaudio, M. Kobayashi, G. Caire, G. Colavolpe, On the effectiveness of OTFS for joint radar parameter estimation and communication. IEEE Trans. Wirel. Commun. **19**, 5951–5965 (2020)

16. R. Xie, D. Hu, K. Luo, T. Jiang, Performance analysis of joint range-velocity estimator With 2D-music in OFDM radar. IEEE Trans. Signal Process. **69**, 4787–4800 (2021)

17. J. Ellinger, Z. Zhang, M. Wicks, Z. Wu, Multi-carrier radar waveforms for communications and detection. IET Radar Sonar Navig. **11**, 444–452 (2016)

18. C. Sturm, W. Wiesbeck, waveform design and signal processing aspects for fusion of wireless communications and radar sensing. Proc. IEEE **99**, 1236–1259 (2011)

19. L. Hu, Z. Du, G. Xue, Radar-communication integration based on OFDM signal, in *Proceedings of the 2014 IEEE International Conference on Signal Processing, Communications and Computing (ICSPCC)*, Guilin, China, 5–8 August 2014; pp. 442–445

20. Y. Xin, C. Huang, Research on FDSS modulation technique. Mob. Commun. **42**, 40–44 (2018)

21. R. Nissel, M. Rupp, Pruned DFT-spread FBMC: low PAPR, low latency. High Spectral Efficiency. IEEE Trans. Commun. **66**, 4811–4825 (2018)

22. D. Na, K. Choi, Low PAPR FBMC. IEEE Trans. Wirel. Commun. **17**, 182–193 (2018)

23. I. Darwazeh, H. Ghannam, T. Xu, The First 15 Years of SEFDM: a brief survey, in *Proceedings of the 2018 11th International Symposium on Communication Systems, Networks and Digital Signal Processing (CSNDSP)*, Budapest, Hungary, 18–20 July 2018; pp. 1–7

24. A. Rashich, A. Kislitsyn, D. Fadeev, T. Ngoc Nguyen, FFT-based trellis receiver for SEFDM signals, in *Proceedings of the 2016 IEEE Global Communications Conference (GLOBECOM)*, Washington, DC, USA, 4–8 December 2016; pp. 1–6
25. S.B. Makarov, M. Liu, A.S. Ovsyannikova, S.V. Zavjalov, I.I. Lavrenyuk, W. Xue, J. Qi, Optimizing the shape of Faster-Than-Nyquist (FTN) signals with the constraint on energy concentration in the occupied frequency bandwidth. IEEE Access **8**, 130082–130093 (2020)
26. J. Liu, W. Liu, X. Hou, Y. Kishiyama, L. Chen, T. Asai, Enhanced non-orthogonal waveform (eNOW) for 5G evolution and 6G, in *Proceedings of the 2021 26th IEEE Asia-Pacific Conference on Communications (APCC)*, Kuala Lumpur, Malaysia, 11–13 October 2021; pp. 56–61
27. Z. Mansor, A. Isa, Performance analysis of power and spectrum efficiency in uplink LTE system. J. Eng. Technol. **1**, 38–45 (2011)
28. A. Şahin, N. Hosseini, H. Jamal, S. Hoque, D.W. Matolak, DFT-spread-OFDM-based chirp transmission. IEEE Commun. Lett. **25**, 902–906 (2021)
29. J. Wang, B. Zhang, P. Lei, Ambiguity function analysis for OFDM radar signals, in *Proceedings of the 2016 CIE International Conference on Radar (RADAR)*, Guangzhou, China, 10–13 October 2016; pp. 1–5
30. I.P. Nasarre, T. Levanen, K. Pajukoski, A. Lehti, E. Tiirola, M. Valkama, Enhanced uplink coverage for 5G NR: frequency-domain spectral shaping with spectral extension. IEEE Open J. Commun. Soc. **2**, 1188–1204 (2021)
31. X. Zhou, C. Wang, R. Tang, Channel estimation based on IOTA filter in OFDM/OQPSK and OFDM/OQAM systems. Appl. Sci. **9**, 1454 (2019)
32. P. Siohan, C. Roche, Cosine-modulated filterbanks based on extended Gaussian functions. IEEE Trans. Signal Process. **48**, 3052–3061 (2000)
33. T. Li, S. Hu, G. Wu, L. Yan, Y. Luo, Power spectral density comparison for the clipped OFDM-type signals, in *Proceedings of the 2010 Second International Conference on Networks Security, Wireless Communications and Trusted Computing*, Wuhan, China, 24–25 April 2010; Volume 1, pp. 269–272
34. R. Chávez-Santiago, M. Szydełko, A. Kliks, F. Foukalas, Y. Haddad, K.E. Nolan, M.Y. Kelly, M.T. Masonta, I. Balasingham, 5G: the convergence of wireless communications. Wirel. Pers. Commun. **83**, 1617–1642 (2015)
35. N. Ye, J. An, J. Yu, Deep-learning-enhanced NOMA transceiver design for massive MTC: challenges, state of the art, and future directions. IEEE Wirel. Commun. **28**, 66–73 (2021)
36. J. Pan, N. Ye, A. Wang, X. Li, A deep learning-aided detection method for FTN-based NOMA. Wirel. Commun. Mob. Comput. **2020**, 5684851 (2020)
37. B. Di, H. Zhang, L. Song, Y. Li, Z. Han, H.V. Poor, Hybrid beamforming for reconfigurable intelligent surface based multi-user communications: achievable rates with limited discrete phase shifts. IEEE J. Sel. Areas Commun. **38**, 1809–1822 (2020)

Chapter 10
Summary and Outlook

This chapter summarizes this book and discusses the future directions of high-frequency wireless communications.

10.1 Summary

In this book, we discuss various aspects to have a thorough view of high-frequency wireless communication technologies. Specifically, in Chaps. 2 to 4, we discuss high-spectrum-efficiency transmission technology towards the three scenarios of efficient system, detection and coding. In Chaps. 5 and 6, we investigate the secure directional modulation in high-frequency wireless communications. In Chaps. 7 to 9, we explore the integrated technology of high-frequency wireless communications and high-precision ranging. The specific contributions of this book are as follows.

1. In Chap. 2, we present an efficient system, the MIMO-NOMA broadcasting network with full-duplex D2D communications, validated through detailed numerical results. Our scheme outperforms conventional TDM-based and LDM schemes without D2D in terms of outage probability and ergodic capacity. Furthermore, we observe that system performance can be enhanced by increasing transmission power, expanding the base station's antenna array, and optimizing the transmission power for full-duplex communications to minimize self-interference and maximize capacity.
2. In Chap. 3, we discuss an efficient detection technology, the multi-user detection algorithm for asynchronous code-domain NOMA, which significantly enhances the traditional factor-graph model by incorporating a time delay dimension. This innovation leads to the construction of a 3D factor-graph model with two message passing mechanisms, effectively characterizing the interference structure in

© The Author(s), under exclusive license to Springer Nature Singapore Pte Ltd. 2025 179
J. Li et al., *Key Technologies of High Frequency Wireless Communications*,
https://doi.org/10.1007/978-981-96-5894-7_10

asynchronous environments. Building upon this model, we developed the low-complexity 3D-EPA to refine interference approximation, ensuring optimal performance. Furthermore, we extended this framework to multi-antenna systems, deriving state evolution to analyze the theoretical BER performance. Our simulation results demonstrate that asynchronous code-domain NOMA, when coupled with 3D-EPA, outperforms its synchronous counterpart, particularly under high overloading conditions. Additionally, the 3D-EPA exhibits linear complexity, comparable to traditional EPA, and we explored its performance across various frame lengths and spreading factors, as well as its convergence behavior.

3. In Chap. 4, we propose an efficient coding technology, the overlap-optimized decoder for high-speed communications, designed to meet the high data rate requirements of sensor networks. By employing a QC-LDPC code that allows up to one-third decoding time overlap, we enhance the decoder's throughput. To balance performance and complexity, a Modified 2-bit MSA is implemented, alongside a Boolean operation-based CNU and VNU, which reduces resource overhead through logic circuit message calculations. A shift-register-based memory strategy addresses the quasi-cyclic characteristic, minimizing read/write latency. The decoder achieves a coding gain of 5 dB at a BER of 10^{-6} and a throughput of 7.76 Gbps at a frequency of 156.25 MHz. FPGA implementation on the Xilinx Virtex UltraScale + FPGA VCU118 board reveals that it utilizes only 10% of the board's resources, underscoring its efficiency and practicality. In summary, this decoder offers a promising solution for high-speed communication applications, particularly in sensor networks.

4. In Chap. 5, we investigate a CE optimal-based hybrid beamforming design algorithm to enhance multi-user physical layer security in communication systems. Initially, we adopt a hybrid beamforming approach, randomly selecting antennas for signal transmission at the symbol rate. However, this method leads to high sidelobe energy, potentially facilitating eavesdropping. To address this, we propose a cross-entropy iteration method to select the optimal antenna combination, thereby reducing the energy available to eavesdroppers and improving system security. Simulation results demonstrate that our proposed method achieves approximately 10 dB lower sidelobe energy compared to random selection and ensures a consistently high symbol error rate of 0.75 for eavesdroppers under QPSK modulation, indicating robust physical layer security performance.

5. In Chap. 6, we study the secure directional modulation utilizing hybrid beamforming in RIS-aided communication networks. We propose a cross-entropy iterative algorithm to optimize the hybrid beamformers at the gNB and the analog beamformer at the RIS, aiming to enhance physical-layer security and reduce sidelobe emissions. Simulation results show a reduction in sidelobe energy by over 8 dB, with eavesdroppers facing symbol error rates of approximately 0.75 for QPSK and 0.875 for 8PSK in undesired directions, thus meeting total security requirements.

6. In Chap. 7, we introduce an enhanced accuracy OFDM radar system that integrates the FRFT and carrier phase analysis techniques. By employing FRFT, we achieve ranging accuracy within one millimeter wave wavelength, further enhanced to 20 μm through phase analysis. This system outperforms existing methods, due to the

parallel processing capabilities of OFDM and the short wavelength of millimeter waves, enabling high accuracy at a high refresh rate over a long unambiguous range.

7. In Chap. 8, we investigate the FTN-enhanced DFT-s-OFDM waveform for ISAC system. By employing DFT spreading and FTN signaling, we achieve reduced PAPR and enhanced spectral efficiency compared to traditional OFDM waveforms used in 4G/5G systems. To further improve system throughput, we propose an FTN-DFT-s-OFDM waveform that incorporates an IOTA filter. The IOTA filter's superior time-frequency focusing characteristics mitigate ISI, thereby enhancing transmission performance. At the receiver end, an FDE receiver is employed for signal demodulation, with a computational complexity of $O(NlogN + N)$. Our simulation results highlight the effectiveness of the proposed scheme, achieving a 3.5 dB reduction in PAPR and a 50% improvement in throughput relative to existing waveforms, while also improving BER performance. These advancements underscore the potential of our waveform design in enhancing the efficiency and performance of communication systems.

8. In Chap. 9, we propose a novel FDSS-enhanced DFT-s-OFDM waveform for ISAC systems, aiming to improve both sensing and communication performance. Our simulations reveal that amplitude fluctuations in the frequency domain of traditional ISAC waveforms lead to noise enhancement, thereby degrading sensing performance. To mitigate this issue, we introduce a framework for FDSS-enhanced DFT-s-OFDM, which utilizes FDSS to achieve a more uniform amplitude distribution in the frequency domain. We design FDSS filters based on pre-equalization and IOTA filters, both of which effectively smooth the frequency domain signals. Simulation results show that the proposed waveform achieves a 4 dB improvement in sensing accuracy and exhibits lower ambiguity function side-lobes, indicating enhanced sensing resolution. Additionally, the FDSS approach significantly reduces the PAPR compared to standard DFT-s-OFDM. These contributions underscore the effectiveness of the FDSS-enhanced DFT-s-OFDM waveform in enhancing the performance of ISAC systems.

10.2 Future Directions

In this section, we discuss the future directions of enhancing high-frequency wireless communications.

1. **High-spectrum-efficiency transmission technology**:
   In the derivation of ergodic capacity and outage probability for the full-duplex D2D assisted high-frequency wireless communication systems, existing researchers adopt some approximations for the sake of simplicity. Although these approximations have a negligible impact on the overall performance, considering theoretical rigor, the approximate methods employed will be subject to further investigation in future work to obtain more precise results.

2. **Secure directional modulation technology**:
   The proposed multi-user low sidelobe directional modulation technique, which generates secure beams based on target user angles, presents a solid foundation for security enhancement. However, the absence of guaranteed security at the target angle necessitates further innovation. Introducing frequency variations to create frequency diversity arrays offers a promising path forward. Such arrays can generate beams that are dependent on both angle and distance, thereby significantly enhancing the security performance of the system.

3. **Integrated technology of communication and sensing**:
   The integrated high-frequency communication and ranging technology, currently implemented with OFDM in a single transmit and receive configuration, demonstrates excellent performance without multi-user interference. To further enhance measurement capabilities, future research should explore multi-transmit and multi-receive configurations for imaging. This approach will not only improve measurement precision but also expand the measurement dimensions. Additionally, addressing the issue of mutual interference among multiple users by analyzing the impact of codewords on interference between multiple radars is essential. Optimizing the codewords for integrated communication and ranging transmission will be key to achieving higher measurement accuracy.

In summary, these future directions aim to refine existing methodologies, enhance security, and advance the integration of communication and ranging technologies. By pursuing these research paths, we can unlock the full potential of high-frequency wireless communications and pave the way for next-generation wireless systems.

The manufacturer's authorised representative in the EU is Springer
Nature Customer Service Centre GmbH, Europaplatz 3, 69115 Heidelberg,
Germany. If you have any concerns regarding our products, please
contact ProductSafety@springernature.com

Printed and bound by CPI Group (UK) Ltd, Croydon, CR0 4YY

01/07/2026

02153166-0002